To Bruce, in

and

appreciation.

Jon Stadtler

SCIENTIFIC METHOD

This book shows how science works, fails to work, or pretends to work, by looking at examples from such diverse fields as physics, biomedicine, psychology, and economics. Social science affects our lives every day through the predictions of experts and the rules and regulations they devise. Sciences like economics, sociology and health are subject to more 'operating limitations' than classical fields like physics or chemistry or biology. Yet, their methods and results must also be judged according to the same scientific standards. Every literate citizen should understand these standards and be able to tell the difference between good science and bad. *Scientific Method* enables readers to develop a critical, informed view of scientific practice by discussing concrete examples of how real scientists have approached the problems of their fields. It is ideal for students and professionals trying to make sense of the role of science in society, and of the meaning, value, and limitations of scientific methodology in the social sciences.

John Staddon is James B. Duke Professor of Psychology, and Professor of Biology and Neurobiology, Emeritus at Duke University. He does research on adaptive behavior, economics, and the history and philosophy of science.

SCIENTIFIC METHOD

How Science Works, Fails to Work,
and Pretends to Work

John Staddon

Routledge
Taylor & Francis Group

NEW YORK AND LONDON

First published 2018
by Routledge
711 Third Avenue, New York, NY 10017

and by Routledge
2 Park Square, Milton Park, Abingdon, Oxon, OX14 4RN

Routledge is an imprint of the Taylor & Francis Group, an informa business

© 2018 Taylor & Francis

The right of John Staddon to be identified as author of this work has been asserted by him in accordance with sections 77 and 78 of the Copyright, Designs and Patents Act 1988.

Trademark notice: Product or corporate names may be trademarks or registered trademarks, and are used only for identification and explanation without intent to infringe.

Library of Congress Cataloging-in-Publication Data
Names: Staddon, J. E. R., author.
Title: Scientific method: how science works, fails to work, and pretends to work / John Staddon.
Description: New York, NY: Routledge, 2018. | Includes bibliographical references and index.
Identifiers: LCCN 2017035127 | ISBN 9781138295353 (hb : alk. paper) | ISBN 9781138295360 (pb : alk. paper) | ISBN 9781315100708 (ebk)
Subjects: LCSH: Science–Methodology. | Social sciences–Methodology.
Classification: LCC Q175 .S7394 2018 | DDC 001.4/2–dc23
LC record available at https://lccn.loc.gov/2017035127

ISBN: 978-1-138-29535-3 (hbk)
ISBN: 978-1-138-29536-0 (pbk)
ISBN: 978-1-315-10070-8 (ebk)

Typeset in Bembo
by Deanta Global Publishing Services, Chennai, India

Better understanding of the scientific method, its strengths, weaknesses, and practical and ethical limitations, can help. The scientific enterprise is larger than it has ever been—there are more living scientists than dead ones. Despite interdisciplinary efforts in some applied areas,[1] science is also more fragmented than ever. Yet there is much need for citizens to understand science (especially as they support much of it!). To understand not just the major facts, but the methods used in different sciences, and their limitations.

Social science, economics especially, affects our lives every day through the predictions of experts and the rules and regulations they have devised. Yet economic 'science' is very different from, and much less secure than, physics or chemistry. The methods of physics also look very different from the methods of social psychology or sociology. But all must be judged by the same standards—standards which should be understood by every literate citizen.

There are two main ways to explain science to a general audience. One is to compile a comprehensive list of scientific methods and analytical techniques as they are used in various disciplines. To teach the rules of the game. This is the tactic of works such as E. Bright Wilson's classic: *An Introduction to Scientific Research.*[2] But such a book will not be read by any but serious aspirants: individuals who want to become scientists or already are scientists. Moreover, listing rules may not in fact be the best way to 'get the idea' of what science is all about. Scientific method is not a set of well-defined rules. Discovery is not an inevitable result when you have ticked to the end of a checklist. The second method, careful study of examples, is the best way to learn about any difficult subject. This book is not a list of rules, but a discussion of examples from physical science, biology and, predominantly, social science.

One anonymous reviewer of an early draft felt that science is now under attack, and anyone who writes about science for a general audience should do his best to defend it. He[3] was unhappy because this book seems "hardly designed to encourage broader acceptance of the legitimacy of the scientific method." That's not the point. Science as an idea should need no defense in a society whose very existence depends upon it. Contemporary science, especially biomedicine and social science, is not in a very healthy state. It needs not cheerleading but understanding—and improvement. My aim is not to disparage scientific methods but to show what they are and how they work in a range of contexts. If science is to thrive, people must understand the difference between good science and bad. They need to be skeptical, to recognize claims that are baseless or exaggerated. They need to understand the limitations of scientific method as well as its strengths.

It should not surprise, therefore, that most of my examples illustrate common flaws or misconceptions. Nor will it amaze that there are more problems in social science—social psychology and economics—than in physics; more in epidemiology and health science than in experimental psychology or chemistry.

I have chosen my examples either to make a point—about the difference between causation and correlation, for example—or because particular papers have been very influential. I look at all the examples from the same perspective: Are experiments involved? Are the results replicable? Do they prove what the experimenters claim? If the data are merely observational, what do they suggest? Are correlations presented as causes? Is there a theory to explain the results? If theory, what kind of theory is it—causal or functional? Can it be tested experimentally or in some other way?

Statistics are involved in much social and biomedical science. Statistics is an area in which I am far from expert, which is an advantage, since most researchers who use inferential statistics aren't experts either. Following the prevailing convention, they treat statistics as a 'black box.' They plug their data in and hope to get an acceptable number out. But the statistical black box is not a passive pass-through. It is an active participant in the work. Hence, it is essential to understand *exactly* what these techniques do, how they work, what they assume, and what they really mean. It is not an advantage to be a mathematical whiz if the task is to explain statistics in a truly comprehensible way. It is essential to begin at the beginning with simple problems and transparent methods. That is what I have tried to do.

Most science involves some kind of theory. Here also there are big differences between physical and biological science and social science. Theories in physics,[4] chemistry and much of biology are about causes and obedience to quantitative laws. Will this chemical react with that? What is the orbit of a newly discovered asteroid? What pathogen is responsible for some new ailment? Not so in economics and some psychology where explanations tend to involve motive and purpose. Outcomes—incentives—are all that matter. Economists usually assume that people act so as to maximize their gains. This is called *rational* behavior. The usual presumption is that there is a single rational solution to every well-defined economic problem. *Rationality* is also treated as a single thing: you're either rational or you're not. But behavioral research suggests that rationality is not a single faculty you can switch on or off. It is one mode of behavior of a complex causal system that almost never pays attention to the marginal utilities of rational economics.

In psychology, outcome-based theories are less common—although the behaviorist idea of *reinforcement* comes close. Instead many theories in social and personality psychology take the form of what one might call *scientization.* They take notions from common speech—folk psychology, if you will—and dress them up as scientific concepts. Sometimes, as with the idea of *intelligence,* this works reasonably well. But at other times, as when self-confidence becomes 'self-efficacy,' empathy becomes 'attribution theory,' and free will becomes a 'self-system,' not so well. I discuss examples of theories from social psychology in later chapters.

Economics takes outcome-based theory to its highest level. Theoretical economists get Nobel prizes for showing that rational behavior leads to predictable market equilibria. Other Nobelists argue that people aren't rational at all. One group shows that markets are 'efficient.' Another that they aren't. Behavior is usually assumed to be utility-maximizing. But more and more apparent exceptions to rational choice have been and are being identified.

What is missing in economics is any serious interest in the *causes* of economic behavior, in its psychology. Economists have been criticized for their myopia in this respect for more than sixty years.[5] Criticism has had little effect. But now, when economists' failures to predict significant economic events have become embarrassing, it is surely time for another look.

The book is not a survey of all science, which would take a lifetime to write and almost as long to read, if it could even be done by one person. The last notable effort along these lines was polymath biologist Lancelot Hogben's splendid 1,100-page *Science for the Citizen* in 1939. Perhaps for lack of space, perhaps for lack of any established principles to expound—perhaps because he didn't like economists—Hogben had little to say about social science. The present volume is much less ambitious than SFTC, but does have more to say about social science. It is about *method*—in observation, in experiment and in theory, especially theory in social science. Some discussion of philosophy and epistemology is unavoidable when theory is the topic but again, discussion of specific cases should help. I have not attempted a comprehensive survey of all the statistical and experimental methods used in science along the lines (for example) of E. B. Wilson's book. My purpose is to help people understand what science *is*, rather than train them to be scientists. So, I have followed Plato's advice: difficult ideas are best grasped through examples.

Notes

1 The subhead of a 2015 *Nature* special issue on interdisciplinarity is 'Scientists must work together to save the world'. www.nature.com/news/why-interdisciplinary-research-matters-1.18370

2 Wilson, E. B. (1952) *An Introduction to Scientific Research.* New York: Dover.

3 I use the masculine generic for simplicity, euphony, and historical continuity: 'he' usually means 'he or she' as it has since the dawn of English.

4 Of course, there are exceptions. String theory in physics gets a lot of criticism for its untestability, for example. 'Dark matter' seems to make up more of the universe than visible matter, yet its origin and even its existence, is still obscure.

5 Edwards, W. (1954) The theory of decision making. *Psychological Bulletin*, 51(4), 380–417.

ACKNOWLEDGMENTS

I thank Nada Ballator, Jeremie Jozefowiez, Max Hocutt, Alex Kacelnik, Peter Killeen, and especially, Kevin Hoover for helpful comments, criticisms and support. I am also grateful to Bruce Caldwell, Roy Weintraub, Neil de Marchi, Marina Bianchi, Richard Hammer, Roni Hirsch, Alex Rosenberg, and the other members of the Center for the History of Political Economy group (HOPE) at Duke for comments and intellectual stimulation over the years.

1
BASIC SCIENCE

There are and can be only two ways of searching into and discovering truth. One flies from the senses and particulars to the most general axioms, and from these principles, the truth of which it takes for settled and immovable, proceeds to judgment and to the discovery of middle axioms. And this way is now in fashion. The other derives axioms from the senses and particulars, rising by a gradual and unbroken ascent, so that it arrives at the most general axioms at last. This is the true way, but as yet untried.

Francis Bacon, comparing deduction and induction as routes to truth[1]

Science is the modern name for what used to be called *natural philosophy*—the study of *nature*, what can be observed by anyone who cares to look. Above all, science deals with what can be *tested by experiment*, although scientific knowledge can also be arrived at in other less conclusive ways. There are parts of the biological and social sciences where testing is difficult. Prehistory must be inferred from what we know and can find now. We can't readily test by experiment an hypothesis about the evolution of the earth. Darwinian evolution was accepted initially because of its explanatory power rather than any direct test. The very many ways in which we can discover and evaluate scientific ideas are collectively termed the *scientific method*.

Some hypotheses are testable, at least in principle, but others are not. The hypothesis that objects of different weights all fall at the same speed is scientific; the idea that there is an afterlife, inaccessible to the living, is not. So-called *private events*, the way you see the color purple, your feeling for your dog, your spouse or your pickup truck, are usually excluded from science, although they may have value for other reasons:

As we have no immediate experience of what other men feel, we can form no idea of the manner in which they are affected, but by conceiving what we ourselves should feel in the like situation.

Thus spake Adam Smith,[2] psychologist, philosopher, exemplar of the Scottish Enlightenment, and acknowledged father of economics, in 1759. Whether you are seeing red or green, even whether you are happy or sad, can be detected by asking and via the state of the brain. But exactly what you experience, so-called *qualia*, the felt quality of private perceptions, cannot be measured or shared. I don't know if 'green' looks the same to you as it does to me. A third party can show that we both see red and green as different, judge red and pink more similar than green and pink, and so on. But I don't know 'what it is like' for you to see green. This has been obvious for a long time,[3] though apparently not to some philosophers. One of the most famous papers in this area, for example, is entitled 'What it is like to be a bat.'[4] Alas, no one knows, or can know. Only the bat knows!

There is of course a human faculty of *empathy*: 'I feel your pain,' and so forth. We think we know how someone else feels, if they behave as we would under like circumstances. But we can't always tell a good actor from the real thing. We don't know; we cannot test the hypothesis. It is not science.

The 'what it is like' trope became for a while a favorite among philosophers. If we can *say* 'what is it like?' then being 'like' something must be something real (maybe) and measurable (maybe not). Many philosophers seem to assume that for every word or phrase there is a real thing, a valid object of scientific study. Unfortunately, as we will see, many social scientists also make this error.

Scientific ideas can come from thoughtful observation—*induction*—or via experiment. Experiments may be designed to test a theory—*hypothetico-deductive*; they may be simply exploratory 'what if?' attempts to satisfy natural curiosity. Or they may be 'thought experiments,' impossible to do in practice but yielding insight in other ways.

Physics abounds in thought-experiments, *gedanken* experiments in German, the language of many physicists a century or two ago. The 'What is it like to be a bat' idea is sometimes presented as a thought experiment. It certainly involves thought, but what is the experiment? Perhaps the most famous *gedanken* experiment is Albert Einstein's musing, at age 16, on what it would be like to ride on a beam of light.[5] What would you see? How fast would the light be going? He thought that the very idea posed problems for the equations proposed by 19th century Scottish genius, James Clerk Maxwell, which were (and, under normal conditions, still are) the standard theory of electromagnetism. After thinking some more, ten years later Einstein came up with the special theory of relativity.

Another physics example is *Schrödinger's cat*. Erwin Schrödinger (1887–1961), an Austrian theoretical physicist, was one of the architects of quantum theory, a theory so mysterious that the most famous quote about it is "If quantum mechanics hasn't profoundly shocked you, you haven't understood it yet," which is usually attributed to the brilliant Danish physicist, Niels Bohr (1885–1962). Schrödinger imagined a cat sealed in a box with a dispenser of poison to be activated by a radioactive source. The uncertainty about the source means, said ES, that the cat is neither dead nor alive until the box is opened.

In its early days, chemistry advanced largely through 'what if?' trial and (sometimes with lethal results) error. Will substance A react with/dissolve/neutralize/explode substance B? The field had to progress in this rather aimless way until enough facts had accumulated that patterns could be discerned and predictions made. Once enough 'pure' substances (elements) had been identified, the Russian, Dmitri Mendeleev (1834–1907) could look at their interactions and arrange them in the periodic table. Gaps in the table in turn suggested the possibility of hitherto undiscovered elements. The table also suggested predictions about how elements would react with one another. This final step is an example of *hypothetico-deductive* experimentation. But the first steps were not hypothetico-deductive. They were simply 'what if?' curiosity-driven attempts.

The hypothetico-deductive experiment asks questions of the form 'If I do X, will I get Y?' where the hypothesis, that X will cause Y, is derived from a theory, past observation, or simply intuition. Hypothesis: Objects of different weights fall at the same rate (Galileo). Experiment: roll 'em down an inclined plane so you can measure the time it takes with 16th century equipment (a water clock). The technique also minimized air resistance, which is presumably what misled father-of-philosophy and teacher of Alexander the Great, Aristotle (384–322BC), to his wrong view: that heavier objects fall faster.

Inductive science infers some general rule from a set of observations—like the periodic table derived from a list of elements and their chemical properties. Or, more simply: all the swans I see are white, *ergo*, swans are all white. As we will see, all these divisions are rather arbitrary. *Scientific method is not an algorithm.* Details matter: the problem to be solved, the question being asked, the nature of the subject matter, all make a difference. This diversity means that scientific method is best understood not as a long list of rules, but through analysis of examples.

Observation

I begin with an example of how science can advance through simple observation. Some six centuries before the Christian era, somebody figured out that the earth is not flat. How did he do it? We don't know for sure, but here is a guess. The diagram shows a method available to anyone close to a cliff on a clear day with a calm sea. The finding is that ships can be seen at a greater distance from the top of the cliff than from the bottom. The diagram shows what this must mean. The horizon is indicated by the two arrows. Given a circular earth, then from the lower vantage point the horizon is obviously closer (A) than from the higher vantage point (B). Many people must have noticed over the years that a distant ship, just visible on the horizon from the higher vantage point, disappears when viewed from the beach.

I used a circle as an example, but obviously this single observation, the higher you are, the more distant the horizon, is consistent with many kinds of curve.

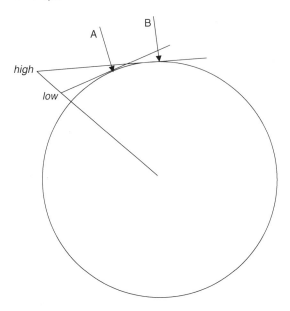

Proving that the earth is more or less spherical took more comprehensive methods.[6] But this single observation suffices to show that the earth is not flat.

I've called this an example of a scientific fact established by observation. But from another point of view, it is an experiment, to answer the question: can I still see the ship on the horizon when I come down from the cliff to the beach? The distinction between observation and experiment is often arbitrary, since observation is rarely completely passive. Even in astronomy, one must pick particular times of day and year and choose to look at one part of the sky rather than another.

Neophobia

It took many centuries for science to be accepted even by civilized (literate, urban) societies. Acceptance is still far from complete. In 2009 only 32% of the nonscientist US public believed in evolution as a natural process. On the other hand, in 2012, 54% of Americans believed in anthropogenic global warming, a much less well-established scientific conclusion. There is even now a semi-serious flat-earth society.[7]

In past times, many people believed that novelty itself is dangerous. Science, which is a generator of new knowledge, has sometimes been unpopular, even in Western societies. The historian of India, Lawrence James, tells a story[8] showing that the subcontent was not always as sympathetic to science as it is now:

> At some date in the 1770s, an English gentleman and child of the Enlightenment presented … a microscope, to a 'liberal minded-Brahmin'

with whom he had become friendly. The Indian peered through the eye-piece at a piece of fruit and was astonished by the 'innumerable animalculae' which he saw. Pleased by his curiosity, the Englishman gave him the microscope. Soon after, the Brahmin destroyed it, telling his friend that what he had seen had left him 'tormented by doubt and perplexed by mystery' to the point where he imagined his 'soul was imperiled.'

Eunuch Chinese Admiral, *Zheng He* made several pioneering voyages of exploration in the 15th century. His fleets involved hundreds of vessels and the largest ships were up to 120 meters in length, many times larger than those of the Portuguese, Spanish, and English explorers of the era. His skill and resources allowed him to voyage as far as East Africa. But, prompted by Confucian bureaucrats, Zheng's fleet was destroyed and his explorations were ended by a new emperor in 1424. The motive was apparently more political[9] than neophobic, but science and innovation were the victims.

The Victorian age in Britain is usually thought of as an engine of creation and very sympathetic to new invention. But by 1934 Sir Josiah Stamp and a handful of others, confronted with "the embarrassing fecundity of technology" proposed "a moratorium on invention and discovery."[10] Groups directly affected by new labor-saving technology have often resisted. Most famously, the Luddites (named after one Ned Ludd, smasher of stocking frames) 19th century textile workers, destroyed weaving machinery in protest at job losses. The general sympathy for novelty, whether science or technology, in most of the developed world is a relative rarity throughout human history. *Neophobia* has been much commoner than *neophilia*.

Induction

The flat-earth experiment is an example of the deductive method. Here is a historical example of the inductive method. In 1854, the city of London suffered an outbreak of cholera, which killed hundreds in a few weeks. What was the cause? At that time, no one really understood how diseases spread. The germ theory of disease had yet to be proposed. The original suggestion came from an unfortunate Hungarian, Ignaz Semmelweis. (Semmelweis behaved very strangely towards the end of his life, dying of causes [syphilis? Alzheimer's?] that are still disputed, at the young age of 47, in 1865.) He noticed that mothers who gave birth attended by physicians were more prone to puerperal fever than mothers attended only by midwives. Physicians, unlike midwives, were often exposed to corpses and other sources of disease. Semmelweis was able to reduce the incidence of fever by introducing hand sanitation. The germ idea only took off after 1860 with Louis Pasteur's study of puerperal fever and other infections. But in 1854 the 'miasma theory' was in fashion. People thought that diseases like cholera were caused by 'noxious exhalations' from swamps and like places.

In those days, there was no domestic water supply in London. People got their water from hand pumps scattered across the city. The pumps had different sources—local wells or piped from the river Thames or one of its tributaries. Dr. John Snow was a physician working in London but was born in the North of England to a modest family in 1813. He did not believe in the miasma theory and sought another explanation for the spread of cholera. He looked at where cases had occurred. He noticed that almost all of them were clustered within walking distance of a particular hand pump in Broad Street (see the image on p. 7). This allowed him to come up with a *hypothesis*, that the water from the Broad Street pump was contaminated in some way that causes cholera.

The hypothesis suggested an obvious test: remove the handle from the Broad Street pump so that no water can be obtained from it. Snow managed to persuade the local council to disable the pump. His hypothesis was confirmed: the incidence of cholera dropped dramatically, proving that the pump was the source of the disease. Modern statistical techniques (more detail will be covered later in the text) had not then been developed, so Snow could not use them. But, like the great majority of scientific pioneers, he did not need them: the effect of removing the pump handle was rapid and dramatic and there was no reason to doubt the uniformity of the subject population—cholera victims. As we will see, statistics are now frequently used to discover very small effects in populations that are not at all uniform, which can lead to trouble. The statistical method has many pitfalls.

Snow's method is not without problems, however. His experiment is what is called an 'AB design': two conditions/treatments are applied in succession: handle/no-handle. In a laboratory context, both conditions would normally be *repeated*, ABAB, just to be sure that B really has the predicted effect. As we will see, *replication* is the best proof in science. In Snow's case repetition seemed unnecessary: the effect of removing the handle was large. ABAB would also have been unethical: restoring the handle might have caused more deaths. In any case his main purpose was to improve public health rather than advance knowledge. Medicine is not science, because its aims are different. The aim of a medical treatment is to cure the patient. The aim of a scientific experiment is to understand the cause of his disease. These two aims—to learn more vs. curing a particular patient—are sometimes in conflict.

Snow himself was aware of the weakness of an AB test. He noticed that the epidemic itself seemed to be in decline: "There is no doubt that the mortality was much diminished ... by the flight of the population, which commenced soon after the outbreak; but the attacks had so far diminished before the use of the water was stopped, that it is impossible to decide whether the well still contained the cholera poison in an active state, or whether, from some cause, the water had become free from it." An ABAB design might have decided the matter. In the end, it was decided by looking for the source of the Broad Street water.

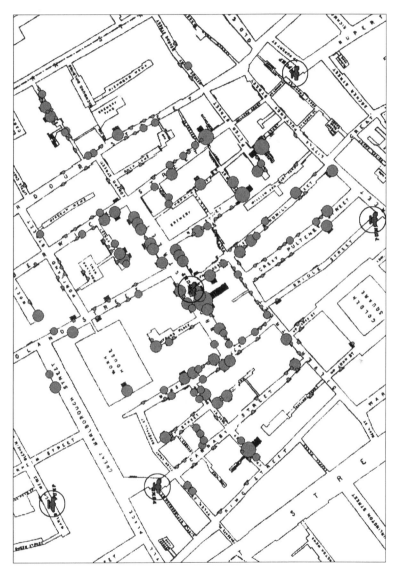

John Snow's map of cholera cases (size of gray circle is proportional to number of cases). A pump is at the center of each circle. Broad Street pump in the center.

Epidemiology

Snow's discovery is considered to be the beginning of the science of *epidemiology*. Now, 'big data' and powerful statistical techniques to sift and sort through them are available and epidemiology has become big business. But the Broad Street example also illustrates a great and oft-forgotten truth: epidemiology is a rich source of *hypotheses* but it cannot prove *causation*. Snow found a *correlation* between the incidence of disease and distance from the Broad Street pump. Proof that the pump was the cause of the disease came from experiment: disabling the pump. Snow's conclusion was correct, even though the experiment was incomplete.

But the cautionary message should be clear: *correlation is not causation*. The inductive method must be combined with the deductive to *prove* a claim. An inductive hypothesis must be tested by experiment, because induction is often wrong: all swans are not in fact white. Unfortunately, science journalism often—usually—ignores this limitation. We are daily bombarded with reports of studies showing a 'link' between this that or the other item of diet or lifestyle and some ailment or other.[11] Most of these 'links' are a waste of file space: because they have not been—and sometimes cannot be—verified by experiment. The next example shows why it is so hard to verify many health claims.

Almost everyone now agrees that smoking cigarettes causes lung cancer. But the path to this conclusion was indirect and a few still question the connection. There are several reasons for doubt. Dr. Snow could be pretty sure that if poisoned water was the problem, eliminating it would show up as reduced cholera incidence with a week or two. But all agree that even if cigarette smoking is a cause of lung cancer and other ailments, its effects are far from immediate. The average age at which lung cancer is diagnosed is seventy. Heavy smoker *Tonight Show* host Johnny Carson died of emphysema in his eightieth year. Delayed onset is also true of the rare cancer mesothelioma, presumed to be caused by exposure to asbestos. Plaintiff's lawyers continue to trawl on TV for mesothelioma victims even though most asbestos use has been banned since 1977.

But in all such cases there *is* a correlation: The chance a smoker in the United States will be diagnosed with lung cancer is about eight times the chance that a comparable nonsmoker will get the disease.[12] At least, those are the most recent statistics. In 1994, however, the relative risk was only two to one. The risk is smaller in some other countries: three to one in China, for example and one to one in some third-world countries although statistics there are less reliable than in the United States and Europe.

There are two reasons for these differences in risk: age of death, and income level. First, you have to live to a good age to die of diseases like lung cancer. If people die at a young age, they are likely to die for other reasons. Hence the risk posed by smoking is less in countries with short life expectancy. Second, poor people tend to die earlier than rich ones. In the 1980s everyone smoked; now smoking is largely confined to the poor; hence, perhaps, the increased correlation between cancer and early death as compared to 1994. Again, this is the problem with correlations: is the cancer risk changing? Or is smoking now confined to a less-healthy population?

How to decide: does smoking cause cancer or not? R. A. Fisher (1890–1962), inventor of the analysis of variance and the randomized-control experimental method, was perhaps the greatest statistician of the 20th century. He was regarded as a prodigy at his elite public (i.e., English private) school, Harrow—other 'Old Harrovians' include Winston Churchill and Jawaharlal Nehru, first Prime Minister of post-imperial India. Fisher also made major contributions to genetics—the 'new synthesis' in evolution[13]—and is regarded by some as the greatest biologist since Darwin. Fisher doubted the cancer-smoking link, for the following reason:

> I remember Professor Udny Yule [also a well-known statistician] in England pointing to [an example] which illustrates my purpose sufficiently well. He said that in the years in which a large number of apples were imported into Great Britain, there were also a large number of divorces. The correlation was large, statistically significant at a high level of significance, unmistakable. But no one, fortunately, drew the conclusion that the apples caused the divorces or that the divorces caused the apples to be imported.

After reviewing the correlational evidence on smoking and cancer, Fisher concluded: "For my part, I think it is more likely that a *common cause* supplies the explanation. [my emphasis]." In other words, Fisher's best guess was that smoking and lung cancer are both caused by some third thing—inflammation, a pre-cancerous condition, whatever—rather than one being the cause of the other.

As in the cholera case, there is an experiment that could settle the issue. Fisher again:

> It is not the fault of Hill or Doll or Hammond [discoverers of the cancer-smoking correlation] that they cannot produce [experimental] evidence in which a thousand children of teen age have been laid under a ban that they shall never smoke, and a thousand more chosen at random from the same age group have been under compulsion to smoke at least thirty cigarettes a day. If that type of experiment could be done, there would be no difficulty [in proving a causal relationship between smoking and disease].

Such an experiment is an obvious and direct test of the smoking-causes-cancer hypothesis. But, for ethical and practical reasons, it is impossible to carry out. The practical reason is the long time-delays involved: the experimenter would have to wait thirty or forty years to accumulate all necessary data. The ethical reason is that no experiment should give people a treatment that is suspected to cause disease.

Indeed, many consider it unethical to *withhold* a treatment that might *cure* a disease, even if it is part of a study designed to improve treatment methods. In 1997, Dr. Marcia Angell, editor of the *New England Journal of Medicine*, wrote in the *Wall Street Journal*:

> Last month I wrote an editorial in the *New England Journal of Medicine* criticizing the National Institutes of Health and the Centers for Disease Control for sponsoring research in Africa in which pregnant women infected with HIV … were treated with placebos [i.e., a neutral—ineffective—drug]. I pointed out that since AZT is known to be highly effective in stopping mother-to-infant transmission of HIV, the researchers are permitting many babies in their care to be born with preventable HIV infection … The NIH and CDC as well as many individual researchers … protested that they were only trying to find shorter, cheaper drug regimens that could be used in Africa, where the full course of AZT is prohibitively expensive… Far from harming African women, they were engaged in a search for an affordable regimen of AZT that would save hundreds of thousands of African children in the future.

Dr. Angell and many others were unpersuaded by the argument of NIH and CDC. Whether you agree or not, this debate just shows how hard it is in practice to do experiments on human beings that have potentially bad consequences for some of the subjects. So, in very many cases we must be satisfied with correlations alone: between dietary fat, or inactivity or low- or high-carbo-hydrate diet, or whatever, and obesity or cancer or high blood pressure or some other measure of ill-health. And we should not be surprised that these correlations reverse from time to time or are inconsistent: fat was bad last year, this year, not so much;[14] sleep too much and its bad for your heart; sleep too little and it's bad for your brain—and so on. Correlation *is not* causation.

Experiment with groups of patients is of course not the last word; it is not the only way to resolve issues like this. Again, Dr. Snow and the cholera epidemic shows the way. Successful as it was, Snow's experiment was incomplete because he did not restore the pump handle to see if cholera returned to the previously affected neighborhoods. Conclusive proof came from another line of research. The Broad Street water came from a well dug just three feet from a cesspit: fecal contamination was the source of the disease. A few years later, research revealed that the toxic bacillus *Vibrio cholerae* is the actual cause of cholera. This is the probing path that must be followed to make sense of many other health-related correlations—diet and disease, smoking and cancer, etc. Are there carcinogens in cigarette smoke? Apparently yes.[15] How do they work? Why are some smokers affected but others not? Until we can identify exactly the physiological and biochemical causes of these afflictions, correlations alone will always leave room for doubt.

Vaccination: A Free-Rider Problem?

Ethics and the limitations on human research are also involved in the practice of mass vaccination. As I mentioned in the Preface, British doctor Andrew Wakefield claimed in 1998 that the MMR (measles, mumps, rubella) vaccine could cause autism in some children. Wakefield's study has been discredited.[16] However, several celebrities, some better informed than others, continue to advocate against vaccination. Many well-informed middle-class parents still don't want their kids vaccinated. They are widely criticized[17] and the scientific consensus is essentially unanimous in favor of vaccination. It is worth noting, therefore, that there is a potentially rational basis for the anti-vaxxers resistance.

The issue is the relative costs of vaccination and nonvaccination. The side effects of vaccination are minimal. Usually the risk of nonvaccination—of getting the disease—is much higher than the risk of vaccination. Nevertheless, the possibility always remains that there is some risk—small but nonzero—associated with any treatment. The vacc/nonvacc logic is as follows:

1 Suppose that the risk to a given child of contracting the disease for which the vaccine is a remedy is p.

2 Probability p depends on the fraction of the population that has already been vaccinated: $p = F(x)$, where x is the fraction of the population already vaccinated and F is a decreasing function of x. In the simplest case, $p = (1-x)$. This is the key assumption: *risk of disease to an unvaccinated individual is inversely related to the proportion of the population that has already been vaccinated*. The more people who have been vaccinated, the smaller the risk to the unvaccinated.

3 Suppose that the cost—to the parents, to the child—of the preventable disease is D, and the cost of the (unlikely but possible) vaccine side-effect is V.

4 Suppose also that the probability of the side effect is q.

Comparing cost of vaccinating vs. not vaccinating yields (assuming that vaccination is 100% effective):

vaccinate: qV
not-vaccinate: $pD = (1-x)D$

As the vaccinated fraction, x, of the population increases, the cost of not vaccinating decreases, perhaps to the point that $qV > pD$, when it would be perfectly rational (in the sense of benefit-maximizing) for a parent to refuse vaccination.

If human parents are guided, as many economists assume, solely by rational self-interest, the fraction, x, of the population that is unvaccinated when $(1-x)D = qV$ represents a stable optimum. The fact that the vaccinated proportion, x, is less than one is nevertheless consistent with maximum benefit (minimum cost) to the community.

Three things make this result problematic, however:

1   Because children are an emotional issue and vaccine cost V is large (autism), parents may well overestimate q and V, hence overestimate the risk of vaccination.

2   Anti-vaxxers may assume that they are in a minority and that most people vaccinate their children, so that the risk of disease, p, is low. These two errors both tend to hold x, the vaccinated fraction, at a too-low, suboptimal level.

3   There is, what economists call, a *free-rider problem*. Although total utility (see Chapter 5 for more on utility) is at a maximum for the population as a whole if the vacc/unvacc ratio matches the marginal costs, as I have described, it is also true that the un-vaccinated are in an ethical sense benefitting at the cost of the vaccinated, who willingly accepted the (small) risk of vaccine side-effect, qV.

This analysis suggests two obvious policies: More research to measure accurately the side effects of vaccination: What are they? How likely are they? And what are the individual differences—are some kids allergic to some vaccines, for example? If the side effects of vaccination are well-known and widely understood, especially if at-risk children can be identified in advance, parents are less likely to misjudge risks. Unfortunately, for the ethical reasons I have just described, because vaccine side effects are likely to be rare, and because many different vaccines would need to be tested, arriving at a clear and undisputed verdict is likely to be difficult. So, the second policy option, which has been followed in many jurisdictions, is to make vaccination mandatory.

John Snow's cholera investigation, like most of science, involved both induction and deduction. Induction, the correlation between disease and distance from the pump, led him to deduction: disabling the pump should halt the epidemic. But there are many other kinds of induction. The paradigmatic case is Darwin and the theory of evolution by natural selection.

Evolution by Variation and Natural Selection

In 1831, a young man of 22 with little scientific education but much interest in natural history, Charles Darwin, embarked on HMS *Beagle*, a small British navy ship charged with hydrographic mapping of South America and whatever other territories its almost equally young Captain Robert FitzRoy could manage. The *Beagle* was just 90 feet long and carried a total of 74 people, including

HMS Beagle in the Straits of Magellan

three natives of Tierra del Fuego, captured on a previous voyage, and now to be returned after being given a 'Christian education' at FitzRoy's own expense. Darwin had a dual role: ship's naturalist and companion to FitzRoy. Captaincy of a small vessel was a lonely occupation in those days. Pringle Stokes, previous captain of the *Beagle*, had committed suicide, so finding a companion for FitzRoy was not just an indulgence. FitzRoy also, despite a successful professional life and pioneering research on meteorology—he coined the word *forecast*—had bouts of depression and himself committed suicide in 1865 after losing most of his fortune.

The *Beagle* voyage lasted five years and took the little ship all around the world. At every landfall, Darwin went ashore and observed the geology and natural history of the place, collecting plants, animals, rocks, and fossils everywhere he went and sending his specimens back to England whenever the ship reached port. His collection methods were opportunistic. He basically tried to get hold of any new thing that struck his fancy. He had no grand theory to guide his collection. His haphazard method turned out to be an asset rather than a liability because it gave him a relatively unbiased sample of the biology and geology of the areas he visited.

He learned that geography is not fixed: the *Beagle* arrived at the port of Concepción in Chile during an earthquake which had raised the land by several feet. Darwin found a layer of sea shells along mountains hundreds of feet above the sea—proof of large topographic changes in the past. He noticed that organisms long isolated from the mainland in the Galapagos islands seem to have diverged from the colonizing species and that living species seem to have similar extinct fossil ancestors. When he eventually put it all together years later, many of the chapter headings in *The Origin of Species* summarize his *Beagle* observations: "On the affinities of extinct species to each other and to living species—On the state of development of ancient forms—On the succession of the same types within the same areas…" and so on. From all his varied observations of geology, zoology, and botany he inferred first (although this was not original with Darwin) that species evolve, and second, his great contribution, that the process by which they do so is natural selection.

Darwin's work is perhaps the most famous example of inductive science. But in this case his theory was accepted not so much because of experimental proof as from its ability to make sense of a vast mass of facts—*empirical data*, in the current jargon.

Experiments to test evolution are hard for two reasons: First, because evolution is usually (but not invariably) slow by the standards of the human lifespan. Second: because it is inherently *unpredictable*. Selection can only work with the pool of genetic and epigenetic *variation* available. Too little is known of how these processes work to allow us to predict with any accuracy what they will offer up over long periods. Artificial selection—breeding farm animals and plants—can produce small phenotypic changes, and sometimes surprisingly large ones as in the case of dog breeds—so the selection process has been experimentally validated. Indeed, the success of artificial selection, by gardeners, pigeon fanciers, and dog breeders, for example, was one of the things that led Darwin to his theory. Nevertheless, uncertainty about the effects of selection over very long periods remains.

Over the years more and more experimental tests of evolution by natural selection have been done. Darwin himself did many experiments with plants. Natural experiments such as the effects of climate variation can also be used as tests. Rosemary and Peter Grant[18] looked at birds on the Galapagos Islands, made famous by Darwin's visit. They studied populations of finches, and noticed surprisingly rapid increases in (for example) beak size from year to year. The cause was weather changes which changed the available food, year-to-year, from easy- to hard-to-crack nuts. An earlier study traced the change in wing color of peppered moths in England from white to black (termed *industrial melanism*) over a few decades, to a darkening of tree bark caused by industrial pollution.[19] As the color of the bark changed, the moths' light camouflage became less effective, darker moths were selected and pretty soon most moths were dark. The modern world continues to battle the continuing evolution of antibiotic-resistant bacteria. The effects of natural selection can be much faster than Darwin thought.

Proof

There is of course an alternative to evolution by natural selection: evolution by *intelligent design* (ID). I will take a little time on this issue because it shows how proof in science, especially on the margins, is not always cut-and-dried. Different people, with different backgrounds, will accept and reject different kinds of proof.

ID is rejected by the vast majority of biological scientists. But it is also important to remember that *consensus is not proof*. Every popular but now discredited theory, from miasma to phlogiston, at one time commanded a majority. Many now-accepted theories—Alfred Wegener's (1912) hypothesis of continental drift, for example—were widely rejected until enough evidence accumulated to overthrow the competition.

"The theory of intelligent design holds that certain features of the universe and of living things are *best explained* by an intelligent cause, not an undirected process such as natural selection[20] [my emphasis]." ID is defended in several ways, from the cosmic (the universe is designed with life in mind) to the molecular. An influential idea is that certain features of organisms are so complex and interdependent there is simply no way they could have evolved by successive small modifications: they are *irreducibly complex*. As Darwin pointed out: "If it could be demonstrated that any complex organ existed which could not possibly have been formed by numerous, successive, slight modifications, my theory would absolutely break down."

Of course, Darwin was being either a little naïve or disingenuous here. He should surely have known that outside the realm of logic, proving a negative, proving that you *can't* do something, is next to impossible. Poverty of imagination is not disproof. In physics, where many laws are established beyond all reasonable doubt, one can say with confidence that a perpetual-motion machine is impossible. Nevertheless, even this most well-established principle, the principle of conservation of energy, still leaves room for doubt, at least for some people. Many don't believe it, so perpetual motion machines continue to be born. On April 17, 2015, for example, a popular news site announced: "Incredible Scientist Makes Free Energy Perpetual Motion Generator." If violations of core principles in physics can still get a hearing, it's hardly surprising that ideas contrary to the consensus in biology still persist.

Darwin was both better informed and less informed than what one might call post-gene biologists. He was in a sense better informed in that the gene had not been discovered (that only really happened with the re-discovery of Gregor Mendel's work in 1900 or so), so he did not rule out heritable effects of the environment. "I think there can be no doubt that use in our domestic animals has strengthened and enlarged certain parts, and disuse diminished them; and that such modifications are inherited ... many animals possess structures which can be best explained by the effects of disuse..." These speculations are incompatible with the idea that only genetic changes are heritable, part of the 'new synthesis' in evolutionary biology, but are now to some extent vindicated by new work on what is called *epigenetics*.[21] Some phenotypic changes can be passed on without genetic alteration.

But Darwin was also less informed about the nature of mutation. He assumed that natural selection worked on changes that were "numerous, successive [and] slight"—even though he knew about so-called 'sports,' which are large changes in phenotype from one generation to the next. In the *Origin* he comments, "I have given in another work a long list of 'sporting plants;' as they are called by gardeners; that is, of plants which have suddenly produced a single bud with a new and sometimes widely different character from that of the other buds on the same plant." Sometimes the change from one generation to the next can be large and even adaptive. So, it is not necessary that Darwinian variation at the phenotypic (whole-organism) level be always small and random. What is conspicuously lacking, however, is a full understanding about the process of variation itself, about its *non*random features. Nevertheless, Darwinian evolution is in many ways even more powerful than Darwin himself thought.

The 'irreducible complexity' defense of intelligent design depends on finding an example so ingenious, so dependent on all its parts being in place before it can function at all, that it cannot have evolved in an incremental Darwinian way. The proponent of this idea gives this example:

> Consider the humble mousetrap… The mousetraps that my family uses in our home to deal with unwelcome rodents consist of a number of parts. There are: (1) a flat wooden platform to act as a base; (2) a metal hammer, which does the actual job of crushing the little mouse; (3) a wire spring with extended ends to press against the platform and the hammer when the trap is charged; (4) a sensitive catch which releases when slight pressure is applied; and (5) a metal bar which holds the hammer back when the trap is charged and connects to the catch. There are also assorted staples and screws to hold the system together.
>
> If any one of the components of the mousetrap (the base, hammer, spring, catch, or holding bar) is removed, then the trap does not function. In other words, the simple little mousetrap has no ability to trap a mouse until several separate parts are all assembled.
>
> Because the mousetrap is necessarily composed of several parts, it is irreducibly complex. Thus, irreducibly complex systems exist.[22]

Is this a valid proof of ID? Well, no one would contend that mousetraps have evolved through variation and natural selection, although one can certainly trace a sort of incremental development for very many technical inventions. Compare modern automobiles to the Model T, or flat-screen monitors to the original tiny cathode-ray-tube devices. Are there systems in the natural world that are not only complex but with such interdependent functioning parts, like the mousetrap, that it is impossible they should have evolved by incremental natural selection?

Finding examples which absolutely elude a Darwinian interpretation is in fact quite difficult—Darwin's review and interpretation of natural history was very thorough. A popular recent candidate is the bacterial flagellum, a rotating tail that propels many bacterial species. This and one or two other such suggestions have in fact been convincingly refuted.[23]

But refutation of individual ID cases is not really necessary, because ID is seriously flawed in another way: what counts as a scientific explanation. The decision to attribute something to 'intelligent design' is both subjective and incomplete. Your intelligent design might be my obvious example of evolution by natural selection. It is incomplete because the cause—intelligent design—is itself unexplained. Unless we assume, as many religious people do, that deistic intelligence is somehow *sui generis*, something obvious and unique, both inexplicable and un-analyzable,[24] ID is not compelling as an explanation of anything. Data alone cannot resolve this conflict.

It is also worth noting that the *variation* part of Darwinian evolution is far from 'random,' even at the gene level. Mutations are not all equiprobable. *Recurrent* mutations turn up more often than others, so the features they control may be quite resistant to selection. Even if mutations are relatively small and random at the genotype level, their effects at the level of the phenotype may not be. The phenotypic effects of mutations are not all small. Change in a single gene can sometimes have large effects: Huntington's disease, cystic fibrosis, and sickle-cell disease are all caused by single-gene mutations. Many single genes also show *pleiotropy*, that is they effect more than one phenotypic characteristic. Single-mutation-caused albinism in pigeons is also associated with defects in vision. Approximately 40% of cats with white fur and blue eyes are deaf; both features are due to the same single-gene mutation. Artificial selection over a fifty-year period for tameness/domesticability in Russian Red Foxes also led to unexpected changes in coat color and other features.[25] There are many other examples.

The point is that *variation*, the process that yields the raw material with which selection must work, is more creative than ID proponents assume. It is also not perfectly understood. The process of *development*, which translates what is inherited into the breeding and behaving adult phenotype, is still the most obscure part of evolutionary biology. The story of Darwinian evolution is far from complete. But ID is very far from an adequate substitute.

Notes

1 Bacon IV [1620/1901], 50: *Novum Organum*, I, Aphorism XIX.
2 *The Theory of the Moral Sentiments* (1759) 6th edition.
3 See, for example, Hayek, F. A. (1952) *The sensory order: An inquiry into the foundations of theoretical psychology*. Chicago, IL: University of Chicago Press, for a useful review.
4 What Is It Like to Be a Bat? by Thomas Nagel appeared in *The Philosophical Review*, October 1974. To be fair to Nagel, he didn't know either.
5 www.nytimes.com/2015/11/01/opinion/sunday/the-light-beam-rider.html?_r=0
6 For example, measuring the curvature directly across a stretch of calm water. Surveyor-trained biologist Alfred Russel Wallace (about whom more later) did just this in 1879 with a six-mile straight stretch of a canal in England called The Bedford Level. https://en.wikipedia.org/wiki/Bedford_Level_experiment

7 www.ncbi.nlm.nih.gov/pmc/articles/PMC4589389/, www.pewforum.org/2009/11/05/public-opinion-on-religion-and-science-in-the-united-states/, http://theflatearth society.org/home/

8 Lawrence, J. (1998) *The Raj: The making and unmaking of British India*. New York: St. Martin's Press, p. 173.

9 www.international.ucla.edu/china/article/10387

10 Merton, R. K. (1973) *The sociology of science: Theoretical and empirical investigations*. Chicago, IL: University of Chicago Press, p. 262.

11 https://opinionator.blogs.nytimes.com/2011/08/09/trying-to-live-forever/

12 These statistics vary widely from source to source.

13 The new synthesis is described in *Evolution: The modern synthesis*, Huxley, J. (1942) London: Allen & Unwin.

14 www.newyorker.com/magazine/2017/04/03/is-fat-killing-you-or-is-sugar

15 https://en.wikipedia.org/wiki/List_of_cigarette_smoke_carcinogens

16 www.ncbi.nlm.nih.gov/pmc/articles/PMC3136032/ or https://en.wikipedia.org/wiki/MMR_vaccine_controversy

17 https://en.wikipedia.org/wiki/Vaccine_controversies, www.nytimes.com/2017/02/08/opinion/how-the-anti-vaxxers-are-winning.html?_r=0

18 Grant, P. R. (1986) *Ecology and evolution of Darwin's Finches*. Princeton, NJ: Princeton University Press.

19 Cook, L. M., Saccheri, I. J. (2013) The peppered moth and industrial melanism: evolution of a natural selection case study. *Heredity*. 110(3), pp. 207–212.

20 www.intelligentdesign.org/

21 www.theatlantic.com/science/archive/2016/11/the-biologists-who-want-to-overhaul-evolution/508712/. See also Staddon (2016), chapter 1.

22 Behe, M. J. *Darwin's Black Box*. [online] Available at: www.arn.org/docs/behe/mb_mm92496.htm

23 See, for example, an article by Eugenie C. Scott and Nicholas J. Matzke: www.pnas.org/content/104/suppl_1/8669.full

24 See *Evolution and the Purposes of Life* for an eloquent defense of purpose in evolution: www.thenewatlantis.com/publications/evolutionandthepurposesoflife

25 https://en.wikipedia.org/wiki/Russian_Domesticated_Red_Fox

2

EXPERIMENT

Then again, to speak of subtlety: I seek out and get together a kind of experiment much subtler and simpler than those which occur accidentally. For I drag into light many things which no one who was not proceeding by a regular and certain way to the discovery of causes would have thought of inquiring after; being indeed in themselves of no great use; which shows that they were not sought for on their own account; but having just the same relation to things and works which the letters of the alphabet have to speech and words—which, though in themselves useless, are the elements of which all discourse is made up.

Francis Bacon, on the experimental method[1]

In the 21st century, science became institutionalized. The number of scientists, especially social scientists, has much increased. The pressure to produce results has increased even more as costs have increased and the number of scientists has grown faster than research support.[2] All has favored a drift towards what might be called the algorithmic approach to scientific method. In biomedicine, for example, the randomized-control-group experiment is often called the 'gold standard' of scientific method. *There is no 'gold standard,'* no single, well-defined method that *guarantees* an advance in understanding. The great advances in science followed a variety of paths, almost none very orderly. As with induction, the experimental method is best understood through examples.

When the phenomenon to be studied can easily be repeated, the single-subject ABAB experimental design can be used. By 'single-subject' I just mean 'no averaging across subjects': the experimenter just looks at one crystal, or one animal, or one human subject[3] at a time. Joseph Priestley's discovery of the gas oxygen is a simple example. Many preliminary studies led to his most famous experiment in 1774. He focused the sun's rays through a large magnifying glass on to a lump of mercuric oxide. The heat caused a gas to be emitted, which Priestley trapped in a glass jar. The gas, later to be called *oxygen*, "was five or six times as good as common air" in sustaining combustion.

To call this experiment 'ABAB' is a bit pedantic. Priestley was simply answering a couple of questions about an easily repeatable phenomenon in the most direct way possible: "If I heat HgO, what do I get?" Answer: gas. "How well can it support combustion?" Answer: better than air. Notice also that the *critical experiment* was the culmination of a long series of preliminary experiments designed to test the effect of heat on different minerals, to collect emitted gases without loss and to find ways to measure the ability of a gas to support combustion.

The experimental method, properly defined, is not any individual experimental procedure or even a list of such procedures. It is not a checklist. It is a *sequence of experiments*—Francis Bacon's *'letters of the alphabet'*—each answering a question prompted in part or entirely by the results of preceding experiments. Science advances if the series ends in some definite, readily testable conclusion about nature. It is the sequence and the conclusion that constitutes the experimental method, not the details of any particular experiment. There is no magic-bullet gold standard that can reveal a great truth in one shot.

The Single-Subject Method

Single-subject ABAB-type experiments are used in those areas of psychology that little depend on a person's history such as sensation and perception: how well people can judge weights or the brightness of a light, or how they respond to visual displays, for example. In the 1830s, the German physician and a founder of experimental psychology, Ernst Weber, studied how well people could discriminate differences in weight. He asked people to judge which weights were heavier; for example, a weight of 1,000gm versus one of 1,100gm, and so on for many different base weights, from 10gm up to 10,000gm. This is a random-order design 'ABBDCBBACABD,' etc. The subject was asked to make many comparisons, so that Weber could get a reliable estimate of what came to be called the *just-noticeable difference* (JND), defined as a stimulus difference judged different 50% of the time. He found that the JND was proportional to the base level: 2gm for a 100gm base and 20 for a 1,000gm base, and so on.

The further assumption that all JNDs produce the same level of sensation, led philosopher-physicist G. T. Fechner to the following logic: JND $= \Delta\psi = k = \Delta S/S$, where $\Delta\psi$ is the constant psychological value of any JND, ΔS is the corresponding physical stimulus change, S is the physical stimulus value, and k is a constant. If $\Delta S/S = k$, then integrating yields $\psi = k\log S + C$, where C is a scale constant: sensation is proportional to the logarithm of the stimulus value. The psychophysical *Weber–Fechner law* holds for many sensory dimensions in addition to weight: brightness, loudness, pressure, electric shock, etc. It is one of the few well-established quantitative principles in psychology.[4]

But psychology is more than psychophysics. In the 1930s many psychologists were eager to understand the process of learning. Unlike sensation, learning is not reversible. Barring transient after-effects, a wavelength of 550nm looks green whether it comes after blue or yellow. But once you have learned something, you cannot easily unlearn it so that the learning experience can be repeated, ABAB fashion, with repeatable results. Therefore, most early learning studies used the 'gold-standard' between-group method—about which more in a moment. However, in 1938 a young assistant professor at the University of Minnesota, after several research years as a Junior Fellow at Harvard, Burrhus Frederic Skinner, proposed a different approach to learning.

Skinner (1904–1990) was for a time America's most famous psychologist. He made it to the cover of *TIME* magazine in 1971 (headline: "We can't afford freedom," the controversial theme of his utopian novel, *Walden 2*). He was interested in how reward and (to a much lesser extent) punishment might be used to change—to *control*—behavior in socially beneficial ways. As was the custom at the time (the 1930s), he began by studying animals,[5] simpler and less ethically problematic than human beings. In a landmark paper called 'A case history in scientific method,'[6] he described how he came up with the two technical innovations that made his name: The *Skinner box* and the *cumulative recorder*.

The method in learning studies at the time was to compare groups of animals, to give one group (the control group) a 'standard' treatment; and apply the treatment to be assessed (size or frequency of reward, degree of food deprivation of the animals etc.) to the other, experimental group. Animals were randomly assigned to each group so that the groups, if not the animals, could be considered as basically identical.

Skinner's experimental method, now called *operant conditioning*, allowed for a simpler approach. To study the effects of intermittent reward he simply trained his hungry animals (usually pigeons) to peck a lighted disk. At first, each peck operated a feeder, via an automatic control circuit, and gave the animals a few seconds access to food. But the birds would continue to peck even if (for example) only every 10th peck operated the feeder (called a *fixed-ratio* schedule of reinforcement), or if only the first peck 60 s after the previous reinforcement operated the feeder (a *fixed-interval* schedule), and so on.

His great discovery was that after sufficient exposure, each of these procedures yields a distinctive pattern of behavior, as revealed by a cumulative record. A great virtue of the cumulative record is that *it is not an average*. (As we will see, averaging can be dangerous unless you know *exactly* how individuals differ.) Every individual peck or lever press and its time of occurrence is represented. Moreover, the pattern for each schedule is usually *stable* in the sense that it can be recovered after exposure to a different schedule. This property of *reversibility* meant that the effects of various procedures could be studied in individual animals, with no need for inferential statistics.[7]

The within-subject method allowed researchers to use the simple ABAB design to investigate the effects of various reinforcement schedules. Many new phenomena were discovered: behavioral contrast, stimulus generalization gradients, the sensory thresholds, surprising perceptual abilities and cognitive skills of animals, the effect of temporal patterns of reward on schedule behavior (the fixed-interval 'scallop,' for example), schedule-induced behavior, laws of choice, the persistent effects of shock-avoidance procedures, and many others. But the contribution of his method that Skinner and his followers thought most important was *control*: control for purposes of education, social melioration, and therapy. They were much less interested in using operant conditioning as a tool to understand the learning process itself.

The single-subject method is limited because although the behavior of an animal on its second exposure to, say, a fixed-interval schedule looks identical to its behavior on first exposure—in an ABAB experiment, behavior under the first B is the same as under the second—although the behavior is the same, the animal is not. Or, to put it more technically, the behavior observed may be the same, but the animal subject is not in the same *state* as before. Because it is not in the same state when it first learns, say, a fixed-interval schedule, it may respond differently to some other procedure after so learning than it would have if the procedure had been applied before any training at all. If our animal could be 'reset' at the end of the experiment, then the effect of procedure C would be the same in an experiment that gave the animal the two treatments AC as in one that gave him ABC. A pigeon may well adapt differently, at least at first, to procedure X after being trained on a fixed-interval schedule than if it had been trained from scratch.

We can't make an ABAB comparison to study learning tasks because we can't 'reset' real organisms. Numerous transfer experiments (i.e., condition A followed by some other condition B) show that every learning experience has some lasting effect. An animal's response to C after AB (ABC)—will often be different than its response to C after A (AC).

Typical fixed-interval (FI) "scallop"

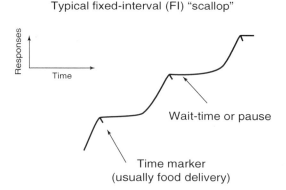

The single-subject method can be used to test some aspects of nonreversible behavior—i.e., learning—but to do so requires hypotheses about the effects of transitions from one schedule to another, about the state of the organism. In other words, it requires a *theory*, something that Skinner strongly discouraged. Skinner's hostility to theory hobbled the development of the field he launched so brilliantly. His emphasis on control and resistance to theory left the operant conditioning rather like the field of animal and plant breeding before the discovery of the gene. Selective breeding worked to produce better horses, dogs, and goldfish, but no one knew why. Applied behavior analysts have a method, but lack full understanding of why it works or, sometimes, fails to work.

The single-subject method cannot answer many obvious learning questions, such as what is the effect of spaced versus massed practice on rate of learning? On retention? On what is learned? Questions like these usually require comparisons between groups of subjects. But going from individuals to groups raises its own set of questions. How well do group-average data represent the behavior of individual subjects? In other words, are all the subjects identical in all essential respects?

The Perils of Averaging: *The Learning Curve*

Suppose, for example, like the followers of Yale experimental psychologist, Clark L. Hull (1884–1967), you are interested in learning and use animals as models. You need to measure how rats learn a simple task, like choosing the correct alley in a maze or pressing a lever for food. You have a group of 20 rats, which you train one at a time, noting after each trial how many of the 20 respond correctly. Initially they know nothing and will fail to press the lever. But over trials, more and more animals will be correct; the *proportion* will increase smoothly from 10% to close to 100%. This *learning curve* can usually be fitted by a number of simple continuous mathematical functions.

The term 'learning curve' has entered common speech. Because learning curves generated in this way are smoothly rising functions, it was assumed for many years that learning in individual animals operates in just this smooth incremental way. Many mathematical learning models were devised based on this assumption. But of course, learning in general is far from incremental. In fact, most subjects in these experiments learn in a step-like way. On trial N, they are at 10%; on trial N+1 and thereafter, they are close to 100%. Where they differ is in the value of N, in *when* they make the step. As the picture shows, this information is obscured by the averaging process. Counterintuitively, the *sudden* learning process yields generally smoother average learning curves than the *smooth*. The reason, of course, is that the *sudden* process never allows performance to decline, whereas *smooth* does.

This is a tricky issue which I will return to later. For the moment, just remember that before averaged data can be interpreted in a meaningful way, we need to know *exactly how* the subjects being averaged differ from one another. In the learning-curve case, for example, the aim is to find the curve form for each

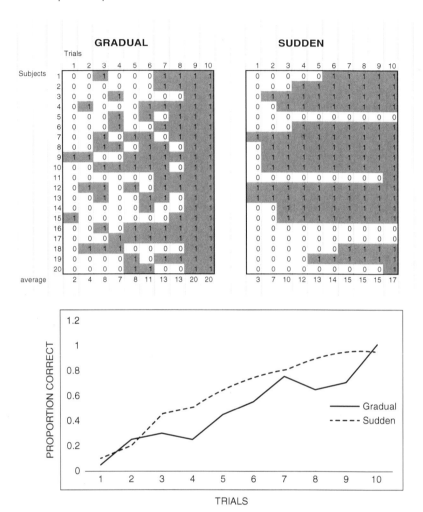

Averaging: What is lost. *Left above, Gradual*: 20 subjects that respond probabilistically correct/incorrect, with the probability increasing linearly from 0.1 to 1.0 over 10 trials. *Right, Sudden*: Subjects that respond correct with probability 0.1, but remember perfectly so never make another error. *Bottom*: A typical graph of average proportion correct trial by trial. The graph for *Sudden* is usually smoother than for *Gradual*.

individual by averaging across individuals and the assumption is that the form is the same for all. Averaging is valid only if in every case the data can be represented as $y_i = F(x_i) + \varepsilon$, where $F(x_i)$ is exactly the same for all individuals, i and ε is a random variable with mean zero (for the learning-curve experiment, y is each animal's choice and x is the trial number). Only in this case will averaging all the y_i yield a valid F.

Closer to reality is the following model: $y_i = F(t_i) + \varepsilon$, where t_i is the time until the subject responds correctly all the time. The Yale psychologists seem to have misjudged both the learning process and the relevant dependent variable. Before you infer anything from an average, you must decide on a model of whatever it is you are averaging. Averages cannot be accepted at face value.

Probability and the Sample Space

When the effect of an experimental treatment is not repeatable, the single-subject, ABAB, method cannot be used. Researchers then must often resort to comparison between groups. Group comparison requires statistics. In all statistical analyses, the first step is coming up with a *probability*—some measure that can tell us how likely it is that the result obtained was accidental, could have come about by chance. This is not a trivial problem, even for very basic experiments. Probability theory uses relatively simple math. But at the same time, it is one of the most conceptually difficult parts of applied mathematics. It is not an accident that three of the founders of statistics as a field, R. A. Fisher, Jerzy Neyman, and Karl Pearson, had a long-running feud over fundamentals[8] even though the fraught issues might seem to be just matters of logic. This sort of thing does not seem to happen in pure mathematics.

There are in fact at least three kinds of probability: (1) relative frequency; this is called *frequentist* probability. This is the meaning I will illustrate with Fisher's tea-lady experiment (below). Probability in this sense is a property of the world, the actual sample space. Bayesian probability, sometimes considered separately, just refers to a particular class of problems, but also requires complete knowledge of the sample space. (2) Probability as *degree of belief*, a psychological category I will say no more about. (3) Chicago economist Frank Knight[9] called it *uncertainty,* and English economist Maynard Keynes called it uninsurable (as opposed to insurable) risk (not that the insurance industry pays much attention to the distinction). Example: in a war, we know that some things will happen that we did not foresee, but we don't know what they are; these are sometimes called 'unknown unknowns.' No number can be *objectively* assigned to uncertainty in this sense.

A sourcebook for the way probability should be used is R. A. Fisher's *Design of Experiments* (1935). Fisher was a man of strong views, and differed in some fundamental ways from his one-time colleague, the older and equally distinguished Karl Pearson.[10] But all can agree on the deceptively simple example with which he begins his book. It shows how probability can be computed exactly in a very English experiment:

THE PRINCIPLES OF EXPERIMENTATION, ILLUSTRATED
BY A PSYCHO-PHYSICAL EXPERIMENT
A LADY declares that by tasting a cup of tea made with milk she can discriminate whether the milk or the tea infusion was first added to the

cup. [How might this be tested?] … Our experiment consists in mixing eight cups of tea, four in one way and four in the other, and presenting them to the subject for judgment in a random order. The subject has been told in advance of what the test will consist, namely that she will be asked to taste eight cups, that these shall be four of each kind, and that they shall be presented to her in a random order … Her task is to divide the eight cups into two sets of four, agreeing, if possible, with the treatments received.

Assuming that the lady gets more cups right than wrong, how can we judge whether she *really* can tell the difference? "In considering the appropriateness of any proposed experimental design, it is always needful to forecast all possible results of the experiment, and to have decided without ambiguity what interpretation shall be placed upon each one of them," wrote Fisher. In other words, the starting point for *all* inferential statistics is to define precisely *all the possible outcomes* of a given experiment, their (frequentist) probabilities and their meaning in relation to the question being asked. Only if this *sample space* is fully defined can we interpret the results of an experiment correctly.

Defining the sample space accurately is often difficult and sometimes impossible. But in the case of the tea-lady experiment, the sample space can be defined precisely. The lady knows that the probability a given cup is tea-first (call it T) is exactly one half (four out of eight). If the successive choices are independent, then her chance of getting the first one correct is ½ getting the first and second correct is just ½ x ½ = ¼ and so on, so that her probability of getting all eight correct is one over 2^8 = 1/256.

But Fisher's estimate is just one in 70, so there's something wrong with that simple analysis. The 1/256 answer would be correct if the experimenter decided on how to mix the tea by tossing a coin each time. But that procedure would not guarantee exactly four T and four M in the eight cups. In other words, the sample space for that experiment is considerably larger than the one Fisher discusses. Hence the probability of getting all correct is much smaller. The correct analysis for the four-out-of-eight experiment…

…may be demonstrated by an argument familiar to students of "permutations and combinations," namely, that if we were to choose the 4 objects in succession we should have successively 8, 7, 6, 5 objects to choose from, and could make our succession of choices in 8 x 7 x 6 x 5, or 1680 ways. But in doing this we have not only chosen every possible set of 4, but every possible set in every possible order; and since 4 objects can be arranged in order in 4 x 3 x 2 x 1, or 24 ways [and order is irrelevant to the task], we may find the number of possible choices by dividing 1680 i.e., by 24. The result, 70, is essential to our interpretation of the experiment.

Because the lady knows that there are exactly four of each type, the number of possibilities is less than if all she knew was that the probability of each cup being T is one half, as in my first analysis. So, even this very simple example can easily be misinterpreted—indeed the *Stanford Encyclopedia of Philosophy*[11] describes a 5-tea-cup experiment with equiprobable outcomes, attributes it to Fisher and gives the analysis with which I began this section.

Fisher goes on to ask just how many choices must the lady get correct if we are to believe her claim? This is the thorny issue of *significance level*. Just how improbable must the experimental result be for us to conclude that our hypothesis—the lady has the ability she claims is true, or at least *not falsified*? Fisher, for practical rather than scientific reasons that I will discuss later, proposed that if we get an experimental result that could occur by chance less than 5% of the time (the $p = 0.05$ level of significance), we should accept it and conclude that we have an effect. For the tea lady, that amounts to getting at least seven correct out of eight.

It is important to understand this very simple example, because it shows the level of understanding needed before statistics can be used in a valid way. The certainty and precision that can be achieved in Fisher's tea-lady experiment is *almost never found* in the statistical procedures that have become standard in much of social, medical, and environmental science. As we will see, it is embarrassingly easy to follow a sort of template that has the appearance of science but which bears the same relation to real science as a picture to its object. American Physicist, Richard Feynman (1918–1988) was a brilliant scientist and a brilliant writer about science (although his words had often to be transcribed by a disciple as he hated writing them down!). He coined the wonderful term 'cargo-cult' science[12] for work that has all the appearances but none of the substance of science, like the wooden planes and ships built by Pacific islanders after the Second World War in the hope of enticing back the Americans who had provided them with goodies during the conflict. Cargo-cult science lives on in much of social science.

Replication and Induction

Can the tea lady *really* tell the difference? We cannot know, because *no* experiment can tell us whether our hypothesis is *true* or not. What it can tell us is whether or not our result is likely to be *replicable*. Says Fisher: "a phenomenon is experimentally demonstrable when we know how to conduct an experiment which will *rarely fail to give us a statistically significant result*. [my emphasis]." Replication is *the* criterion for truth, for proof of causation, in science. Economists, who rarely do experiments, spend much time discussing causation. But causation— replication—*is* an experimental concept. In the absence of experiment, the ability to manipulate, to present or not, a putative cause, any conclusion is hypothetical.

So, is replication the true 'gold standard'? Well, no—because there *is* no such standard—but it *is the best we have*. We may repeat an experiment and get the same result as many times as we like and yet, as philosophers have pointed out, there can never be any *guarantee* that our next attempt will also be successful. No guarantee that has the force of, say, 2 x 2 = 4 or Pythagoras' theorem. The peerless Scot, David Hume (1711–1776), most penetrating of Enlightenment thinkers, pointed out that our belief in *induction*—what has happened in the past will continue to happen—is entirely a matter of faith. It cannot itself be proven:

> It is impossible, therefore, that any arguments from experience can prove this resemblance of the past to the future; since all these arguments are founded on the supposition of that resemblance. Let the course of things be allowed hitherto ever so regular; that alone, without some new argument or inference, proves not that, for the future, it will continue so.[13]

Easily disproved inductive claims—*all swans are white*—show that Hume was right to question the reliability of induction. But it is important to realize that apparently much sounder claims: "I have done this experiment ten times always with the same result, hence my hypothesis is true," are also provisional and just as epistemologically weak. Your inference may be wrong— some unrecognized variable may be responsible for the result. So, someone else's effort to replicate, using an apparently similar procedure that lacks the critical variable, may fail.

Even if both procedures are really identical, the laws of nature may have changed. Surely, we can't question that, can we? Well, yes, we can; we have no *proof* that the universe will stay the same from day to day. Which is fine, actually, because it surely will. But it is as well to remember that this belief is also a matter of faith. Faith in the invariance of natural laws is essential to science. But it *is* faith, just like belief in transubstantiation or the afterlife.

Nevertheless, it is a mistake to be too critical. Without some faith in induction much of science—and essentially all of predictive economics—could not proceed at all. But there is an *order of believability*, the likelihood that we have proven a real causal relation: long-established scientific laws > replicable experiment > reliable correlations > other results from induction.

Notes

1 Bacon, F. *Instauratio Magna*, Distributio Operis: *Works*, vol. 4, p. 24.
2 https://nexus.od.nih.gov/all/2014/03/05/comparing-success-award-funding-rates/
3 I use the word 'subject' to describe the individual entity that is the subject of research, be it a person, an animal, or an object. Human experimental subjects are sometimes called 'participants.'
4 Staddon, J. E. R. (1978) Theory of behavioral power functions. *Psychological Review*, 85(4), pp. 305–320. [online] Available at: http://hdl.handle.net/10161/6003
5 Pedants will note that humans are animals too. But rather than add 'infrahuman' to every animal reference, I rely on the common sense of the reader.
6 Skinner, B. F. (1956) A case history in scientific method. *American Psychologist*, 11(5), pp. 221–233.

7　Murray Sidman's *Tactics of scientific research: Evaluating experimental data in psychology.* New York: Basic Books, 1960, is an important review of the within-subject method.

8　Lenhard, J. (2006) Models and statistical inference: The controversy between Fisher and Neyman–Pearson. *The British Journal for the Philosophy of Science*, 57(1), pp. 69–91.

9　https://en.wikipedia.org/wiki/Knightian_uncertainty

10　Pearson, K. and Fisher, R. A. (1994) On statistical tests: A 1935 exchange from *Nature* Author(s): Karl Pearson, R. A. Fisher, Henry F. Inman. Source: *The American Statistician*, 48(1), pp. 2–11 Published by: American Statistical Association Stable. www.jstor.org/stable/2685077

11　https://plato.stanford.edu/entries/statistics/

12　This is from Feynman's famous speech at Caltech. Available at: http://calteches.library.caltech.edu/51/2/CargoCult.pdf

13　Hume, D. (1975) *Enquiries concerning human understanding and concerning the principles of morals.* 1777. Edited by PH Nidditch.

3

NULL HYPOTHESIS
STATISTICAL TESTING

The first object of statistics is to construct a hypothetical infinite population, of which the actual data are regarded as constituting a random sample.

R. A. Fisher

Le savant doit ordonner; on fait la science avec des faits comme une maison avec des pierres; mais une accumulation de faits n'est pas une science qu'un tas de pierres n'est une maison.[1]

Henri Poincaré

The data-analysis method you should use depends on what you want to know. Are you interested in a specific effect for its own sake, or as a guide to future research? As the earlier examples showed, most experiments in a successful research program are of the second kind: guides to future experiment and/or theoretical development, rather than endpoints. In experimental psychology, for example, several steps were necessary to arrive at the now widely used technology of operant conditioning.[2] The first step was to run rats in a semi-automated maze; in the next development, the animal did not need to navigate a runway but could depress a lever that produced food. This step, allowing the animal to make an arbitrary response to get food, had also been taken by a British researcher, G. C. Grindley, who used head-turning as the response and guinea pigs as the subject. But Grindley's guinea pigs were restrained. Skinner showed that restraint was unnecessary—the same orderly patterns could be obtained from freely moving animals. Skinner arrived at the final procedure via a series of informal, unpublished, and probably unpublishable, stages. Once the technology had been developed it could be used by Skinner and many others to make the discoveries briefly described in Chapter 2.

None of these steps required statistics because all the effects were repeatable with individual subjects. But group experiments also can be divided according to whether the result is an endpoint or a signpost to guide future work. Here is an example that has both features: the result is of value but it may also highlight a direction for future research. The issue was pollution. Birds eat bugs which live in polluted water. What will be the effects of these pollutants on the birds? Since the effects of pollutants are likely to be long-lasting if not permanent, the problem cannot be studied in individuals: a group experiment was necessary. Since the effects of a pollutant are likely to vary only quantitatively from subject to subject, averaging across subjects is legitimate. The authors write:[3]

> Here, we show for the first time that birds foraging on invertebrates contaminated with environmental pollutants, show marked changes in both brain and behavior … male European starlings (*Sturnus vulgaris*) exposed to environmentally relevant levels of synthetic and natural estrogen mimics [EDCs] developed longer and more complex songs compared to control males, a sexually selected trait important in attracting females for reproduction. Moreover, females preferred the song of males which had higher pollutant exposure, despite the fact that experimentally dosed males showed reduced immune function.

This result, if it is real, obviously has implications for environmental regulation and for avian biology. The biological implication concerns the apparently contradictory effects of these compounds: they enhance Darwinian fitness by making the males more sexually attractive; but they also reduce it by diminishing their immune response. Which effect is stronger? Additional experiments could decide. The results would be interesting both for biologists and environment agencies.

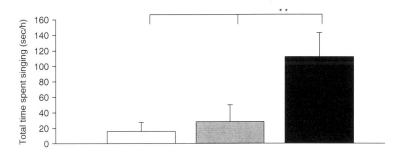

Total time spent singing (sec/h) of male starlings in three treatment groups: control (open bar); E2 dosed (hatched bar); and the chemical mixture dosed (black bar).

Source: Markman S, Leitner S, Catchpole C, Barnsley S, Müller CT, Pascoe D, et al. (2008) Pollutants increase song complexity and the volume of the brain area HVC in a songbird. *PLoS ONE* 3(2): e1674. doi:10.1371/journal.pone.0001674

The data are statistical; the measure in the picture is the total time each group of 12 birds spent singing. Look at two bars at either end in the picture. The first bar is the control (untreated) group the last one is results from the group that got the chemical mixture under test. The short lines at the end of each bar are half the standard deviation (a simple measure of variability) for each group. The two ★★ indicate that the difference between the end bar and the control was statistically 'significant at $p = 0.007$,' i.e., much better than the traditional 0.05 criterion.

Great! The authors have a highly (by the usual standards) significant result. But in order to get an idea of its implications for policy, we also need to know how big the effect is. After all, if the 'significant' effect were tiny—just a 1% change in the dependent variable (singing time), for example, it would be hard to see any need to regulate environmental EDCs. So, knowing the *effect size* is vital if the research result is to be used to guide public action.

Effect size is not an exact measure like mean or standard deviation. It can in fact be measured in many ways. The simplest is probably the best. Looking at the picture you can see that the difference between the total time spent singing for the control and experimental (right-hand bar) groups is about 91 s, or an increase of about six times for the drugged birds. That seems like a big effect, but we must also consider the *variability* of the two groups: a standard deviation of about 60. A large difference between two highly variable groups is less reliable and less important, than the same difference between less variable groups. Dividing the difference in means by the pooled variability of the two groups gives a more balanced measure. In this experiment, it yields an effect size of 1.5 or so, which is generally considered (yes, we're back in the realm of subjectivity again) quite large. Thus, this result might be a useful guide for environmental regulators.

It is a sign of the scientific *zeitgeist* that these investigators, distinguished behavioral ecologists all, even though they found some pretty big effects, make no mention of effect size. All emphasis is on statistical significance.

Statistical Significance

Effect size and statistical significance are imperfectly related to one another. The greater the variability of measures, the less likely they are to be significantly different. It all depends on the sample size. A tiny correlation, like 0.01—which would be invisible in a scatter plot—may be statistically significant with a large enough sample of 1000 or so. But such a small correlation would be truly useless as a guide to policy. Indeed, to act on it might be dangerous, because costs would be likely to outweigh any benefits. *Significance level* is grossly misnamed. By itself statistical significance is of almost no value as a practical guide.

On the other hand, if—and it is a very big if—we are testing a precise and well-supported theory that predicts a tiny difference between two conditions, then significance level may be a useful test. But theories with the required precision and empirical base are nonexistent in the social sciences. It follows that the almost total reliance on statistical significance as a criterion for validity is utterly misguided.

Just how improbable must a statistical result be before we reject the 'no effect' null hypothesis—before we conclude there *is* some effect? The level of improbability—the criterion significance level—we choose is in fact completely arbitrary. In physics, statistics are occasionally used as part of testing a theory. The published significance levels tend to be exceedingly small. A test of gravitational-wave theory[4] in 2016 for example, reported results significant at a probability of one in *3.5 million*—a lottery-level chance that the result was accidental. Social science is much more relaxed.

How are significance levels, on which so much reliance is placed, actually measured? To get an accurate estimate of the chance that a hypothesis is true, in the sense of replicable, the sample space must be known exactly. As in the tea-lady experiment, we need to know all possible outcomes of the experiment or of the sampled dataset, along with their probabilities. As I will show in a moment, doing this properly may require massive repeats of the null (= no-manipulation) condition—which is rarely practicable.

I begin with the tea-lady experiment. Let's look again at the probabilities. For simplicity, let's assume the T- versus M-first assignment is random with p = 0.5, for each cup. This yields, as I showed earlier, 256 possible sequences of eight choices. Just one of these sequences is 'all correct.' So, how many errors will permit us to still decide: "yes" the lady has the talent she claims?

The table shows a few possibilities. The left column is the number she gets correct; the right column the probability of getting that number if she is just guessing. 50:50 guessing is the *null hypothesis* here:

# correct out of 8	probability p
8	0.0039
7	0.03
6	0.11
5	0.21

Just how good a score must she get before we accept her claim: 8, (p = .0039) 7, 6 (p = .11)? There is no absolute truth here. Picking a probability is completely arbitrary. R. A. Fisher suggested that we should reject the null hypothesis if the lady scored a probability of 5% (p = .05) or less. In other words, we will reject the 'random guessing' hypothesis if she gets 7 or 8 right.

Why p = 0.05?

Why 0.05, rather than 0.1 or 0.001? There are socio-economic reasons 5% has been popular for many years (more on that later). But in Fisher's case, the reason may be the kind of research he was doing at the time. In the 1920s, he worked in the United Kingdom at an agricultural research center, the Rothamsted Experimental Station,[5] doing statistics on crop yields. The aim of the research was to compare 'differences in fertility between plots' and measure the contribution to fertility of 'different manures, treatments, varieties, etc.' In other words, the aim was not to understand a process—how fertilizer works or how varieties differ biochemically. The research was not basic—to understand—but applied: to solve a practical problem. There was a need to make a decision about how resources—fertilizer, crop varieties, etc.—should be allocated for maximum yield. *A choice had to be made*, the resources had to be allocated somehow.

But in basic research the aim is to understand, not to act. There is *no need to come to a definite conclusion* if the data do not warrant it. Hence, significance level is simply an inappropriate way to evaluate basic research data. Size matters a lot: a big effect points more decisively to a conclusion, or to the next experiment, than a small one, no matter how statistically 'significant' the small one may be. Yet effect size is often ignored, as in the songbird example.

The 5% significance level seems reasonable if we must allocate our resources somehow and must make some decision. But notice, to do the job properly Fisher would have had to weigh the costs of errors against the benefits of a given allocation. If the costs of error turn out to be high relative to the benefits of getting it correct, we will be reluctant to make a mistake, so we will set the *p* value very low.

For example, let us suppose that Fisher is comparing two fertilizers, A and B, with costs per acre of c_A and c_B. In a test, A is better than B at the 5% level; but A costs (for example) five times as much as B: $c_A/c_B = 5$. Then the estimated yield from A must exceed the estimated yield from B by an amount sufficient to outweigh the greater cost of A. Both significance level and effect size are important in making this judgment. On the other hand, if the cost of fertilizer is a negligible part of the total farm budget, then significance level alone may be sufficient. This kind of analysis can guide choice of significance level in an applied context. But, if the aim is basic science, the acceptable significance level should probably be much lower than 5%. In basic research, a conclusion should never rely solely on significance level.

Unfortunately, weighing costs and benefits was not part of Fisher's analysis, even though his rationale for choosing significance level seems to demand it. Perhaps if he had completed the project by doing a cost–benefit analysis, later researchers would not have been tempted to use the 5% significance level as the standard way to validate statistical comparisons.

Errors

There are two kinds of mistake we can make in any statistical comparison. We can miss a 'true'[6] conclusion or we can accept a false one. In practice, it works like this.

We begin with the null hypothesis, which is what we, the researcher, thinks/hopes is *not* true. Opposed to the null is the often undefined experimental hypothesis (also called the *alternate hypothesis)*, which is what we think/hope *is* true. As we will see, this division is a source of much confusion. The null may be false, but that tells us little about what might actually be true. Researchers are almost invariably content with a significance level that allows them to reject the null hypothesis as not true, while remaining silent on what, exactly, they think *is* true.

Here are the two types of mistake a statistical researcher can make:

> **Type I error:** rejecting the null hypothesis when it is true
> **Type II error:** failing to reject the null hypothesis when it is false

These two errors look pretty symmetrical (although Type II involves a double-negative, which is always confusing). But they are not. Back to the tea-lady experiment: Type 1 is set by the significance level. If we decide to reject the null hypothesis and conclude that the lady has some ability to tell which is which, for example, at the 5% level, then, if in fact she is just guessing, we will be wrong and reject the null hypothesis just 5% of the time.

But Type II error is quite different. The null hypothesis—the lady is guessing ($p = 0.5$)—is perfectly well defined, so estimating Type I is straightforward. If we choose the 5% level, then getting seven or eight correct will cause us to reject the null. But that well-defined null hypothesis *can be false in many ways*. The Type II error depends on just which way. In other words, it depends on an often unspecified *alternate hypothesis*, which you, the experimenter, must devise.

If we have chosen to reject the null when the lady gets seven or eight right, then we will fail to reject it when she gets fewer correct. But that will mean that if she in fact is able to choose correctly 0.6 of the time (i.e., not completely 50:50), we will fail to reject the null .894 (= 1 − .0168−.0896) of the time, an error probability of almost 90%, for this alternate hypothesis, a large Type II error. The probabilities of the relevant outcomes are compared in the next table:

Choice prob.:	*0.50%*	*0.60%*
# *correct out of 8*	*probability p*	
8	0.0039	0.0168
7	0.03	0.0896
6	0.11	0.209
5	0.21	0.279

This pattern of results—failure to reject the null hypothesis when the alternate hypothesis is slightly different from 50:50 randomness—reflects lack of *power*. A sample of just eight choices is too small to pick up small deviations from 50:50. More choices, a larger sample, means more power. For example, if Fisher had used only two cups of tea, 50:50 milk-first, the best the tea lady could do would be to get both right. The chance of that is just ½, × ½ = 1/4. Doing a two-cup experiment obviously gives us little confidence in our ability to judge her abilities. The design is not powerful enough. Eight cups, four of each, is obviously better, but still perfect performance could occur by chance one in 70 times. Purely random equiprobable assignment of T and M is even better. The chance of getting them all right by accident drops to 1/256. If you have only got eight cups, that would then be the best, most powerful design. So, a bigger group, a larger sample space, is generally better. But as we will see, size is no panacea.

Causation

The starling study was an experiment. The effect it found was large and is indeed causal: EDCs affect song production in starlings. Many areas of science deal not with experimental manipulation but simply compare groups. The groups may be people or genomes or drugs. Correlation—the terms 'link' and 'association' are also used—is the method of epidemiology:[7] is there a correlation between this or that element of diet or environment and this or that disease? It is also the method of much biomedical research, looking for biomarkers correlated with disease or some physiological characteristic. Social scientists also do correlation: for example, looking to see if there is a correlation between zip code or race or age and income, health, or longevity. But correlation is not the same as cause.

The widely misused term *link* is often used as if it means *cause*—and not just by journalists, but by supposed 'experts' who should know better. For example, here is a study[8] from Harvard's T. H. Chan School of Public Health which apparently showed that "consuming a plant-based diet—especially one rich in high-quality plant foods such as whole grains, fruits, vegetables, nuts, and legumes—*is linked with* substantially lower risk of developing type 2 diabetes [my emphasis]." This description of a link is immediately followed by a policy recommendation. Recommending action assumes that the link is in fact a cause: a veggie diet *will* make you healthier: "This study highlights that even moderate dietary changes in the direction of a healthful plant-based diet *can play* a significant role in the prevention of type 2 diabetes [my emphasis.]" That's "can play" not the more accurate "might play."

What the study actually found was not a causal relation but a modest *correlation*:

> [R]esearchers followed more than 200,000 male and female health professionals across the U.S. for more than 20 years who had regularly filled out questionnaires on their diet, lifestyle, medical history, and new disease diagnoses as part of three large long-term studies. The researchers evaluated participants' diets using a plant-based diet index in which they assigned plant-derived foods higher scores and animal-derived foods lower scores ... The study found that high adherence to a plant-based diet that was low

in animal foods was associated with a 20% reduced risk of type 2 diabetes compared with low adherence to such a diet. Eating a healthy version of a plant-based diet was linked with a 34% lower diabetes risk, while a less healthy version—including foods such as refined grains, potatoes, and sugar-sweetened beverages—was linked with a 16% increased risk.

Many subjects—200,000—were involved, meaning they had to rely on self-reports of diet, a notoriously unreliable method. The large size of the sample is both good and, possibly, bad depending on what you want to do with the data. Large size means that quite small effects will be statistically significant: "extremely large studies may be more likely to find a formally statistical significant difference for a trivial effect that is not really meaningfully different from the null." Or, more forcefully, "A little thought reveals a fact widely understood among statisticians: The null hypothesis, taken literally (and that's the only way you can take it in formal hypothesis testing), is always false in the real world… If it is false, even to a tiny degree, it must be the case that a large enough sample will produce a significant result and lead to its rejection."[9]

Here is another large-sample study[10] (6.6 million people) that underlines the point. This Canadian study was published in *The Lancet* and internationally reported. Its thirteen[11] authors found another statistically significant, quantitatively trivial and causally completely unproven link—this time between living next to a busy road and developing Alzheimer's disease.

The effects in these studies are relatively modest: 7% increased risk of dementia (the phrase 'risk of' itself implies—quite unjustifiably—a causal relation) if you live within 50 meters of a busy road which shows again how the power of a large sample to seem to justify a tiny effect. The 34% lower diabetes risk the Harvard researchers found is larger. But contrast it with the eight-times-higher lung-cancer rates of heavy cigarette smokers that was necessary to convince us of the dangers of cigarette smoking.

Causation was not, indeed could not, be proved in either case, diet or dementia. All these studies can legitimately do is suggest a *hypothesis*, that a veggie diet *may* lower the risk of diabetes, or that it is better to live far away from a highway. An experiment, or fundamental understanding of brain function and the metabolic roles of vegetable and animal foods, is necessary to establish a cause. We saw earlier the ethical and practical problems doing health-related experiments with human subjects. Harvard's T. H. Chan School is unlikely to do the experiments that would be necessary to warrant its dietary health claims. But that is no excuse for going publicly so far beyond what this study, and many others like it, can actually prove. It is perhaps excusable for a journalist to confound cause and correlation, especially if the topic is one like dementia that induces both fear and curiosity in his readers. But it is irresponsible for the Harvard School of Public Health to confuse the public in this way. And a study with 200,000 subjects is not cheap!

Correlational research cannot establish *causation*. Ill health may be associated with certain zip codes, but it is unlikely that the code, or even the physical environment it represents, is the real cause of any differences that are detected. Occasionally, as in the case of the smoking-cancer link,[12] the strength of a correlation and its general plausibility (not to mention politics!) may bring us around to assuming a causal relation. But only experiment, or new science beyond the level of group statistics—like going from Dr. Snow's correlation with the Broad Street pump to the detection of polluted water and the bacterial cholera agent—can *prove* causation beyond a reasonable doubt.

Experiments with groups of subjects require statistical tests to evaluate their results. The statistics are the same as those used to compare groups of any kind, including those derived without the aid of experiment. The standard of significance is generous: results that are much more likely than one in a million to arise by chance are routinely accepted as significant, indeed they are accepted as *fact*. Social science, pharmacology, psychobiology: all accept Fisher's 5%. But think for a moment what this means for the Type I error rate. If, for example, a thousand identical experiments were done, all testing for a nonexistent effect (ineffective drugs, for example), 50 or so would be 'significant.' Many would be published and most would be unreplicable. The flaws of this much-too-forgiving level have recently become apparent and the method of *null hypothesis statistical test* (NHST) is currently undergoing something of a reform if not a revolution.

Bad Incentives, Bad Results

> *The persistence of poor methods results partly from incentives that favour them, leading to the natural selection of bad science ... Some normative methods of analysis have almost certainly been selected to further publication instead of discovery.*[13]

Despite these uncertainties, NHST is still widely used in social science. The standard method involves an experiment with two (or more) groups. As in the starling study, the groups are matched as far as possible on every relevant characteristic by assigning subjects randomly. The size of the groups varies from study to study, from as small as ten to several hundred in large experiments. One of the groups is defined as the *control group*, which does not receive the experimental manipulation, whatever it may be. The other is the *experimental group* which receives the treatment of interest. The hypothesis to be tested, the null hypothesis, is that the two groups are the *same*—that any measured difference between them could have arisen by chance. Thousands of published studies follow this basic paradigm.

This simple method, so popular in social psychology, offers many opportunities for error and even fraud. In the first decade of the millennium, for example, an energetic young Dutch researcher named Diederik Stapel published numerous social psychology studies showing that:

[W]hite commuters at the Utrecht railway station tended to sit further away from visible minorities when the station was dirty … that white people were more likely to give negative answers on a quiz about minorities if they were interviewed on a dirty street, rather than a clean one … that beauty product advertisements, regardless of context, prompted women to think about themselves more negatively, and that judges who had been primed to think about concepts of impartial justice were less likely to make racially motivated decisions.[14]

It turned out that Professor Stapel was in fact a massive fraudster and most if not all these 'studies' were phony. But Professor Stapel had an eye and ear for what would draw media attention. The topics he chose and the 'results' he purported to get are exactly the sort of content that finds its way on to radio, TV, and the internet.

Despite the temptations—which can be great if, like many biomedical researchers, your progress-dependent, three- or five-year-renewable research grant is your chief source of income—fraud is not the problem with most NHST research. The following quote, from an eminent food scientist in the SC Johnson School of Business at Cornell University and reported in an article called *Spoiled Science* in the *Chronicle of Higher Education*, is a symptom of the prevailing malady:

[Professor Brian] Wansink and his fellow researchers had spent a month gathering information about the feelings and behavior of diners at an Italian buffet restaurant. Unfortunately, their results didn't support the original hypothesis. "This cost us a lot of time and our own money to collect," Wansink recalled telling the graduate student. "*There's got to be something here we can salvage.*"[15]

[my italics]

Four publications emerged from the 'salvaged' buffet study. The research topic, attitudes in an Italian restaurant, is of practical interest to restaurateurs but unlikely to shed light on fundamentals such as the nature of human feeding behavior. The study is correlational not causal—no experiments were done. The topic is entertaining. These are all characteristics typical of most of the 'science' you will read about in the media: a distraction and a waste of resources, perhaps, but not too harmful. The bad bit is in italics. It is hard to avoid the impression that Professor Wansink's aim is not the advancement of understanding, but the *production of publications*. By this measure, his research group is exceedingly successful: 178 peer-reviewed journal articles, 10 books, and 44 book chapters in 2014 alone. Pretty good for ten faculty, eleven postdocs and eight graduate students.

The drive to publish is not restricted to Professor Wansink. It is almost universal in academic science, especially among young researchers seeking promotion and research grants. The concept of the LPU[16] has been a joke among researchers for many years. A new industry of 'pop-up' journals, often with completely fake credentials, has arisen[17] to ensure that no fledgling manuscript fails to find a citable home. I get a solicitation in my inbox almost every week from a new pop-up.[18]

These pressures all favor a generous criterion for statistical significance. If 5% will get any eager scientist a publication after a reasonable amount of work, even if there is actually no real effect in any study, the forces maintaining that very relaxed level may be irresistible.

Research aimed simply at publication is unlikely to contribute much to science. Why? Because, science aims at *understanding,* not any tangible product. Happy accidents apart, most successful research inevitably involves many incomplete tries, many failed experiments, many blind alleys, much activity that is unpublished or unpublishable. It is trial and error, with emphasis on the *error.* Real science rarely results in a long publication list. In social science, especially, such a list is likely to contain more noise than signal.

Darwin and Wallace

A well-known story illustrates one aspect of the problem. In 1857 Charles Darwin received a letter written by the younger Alfred Russel Wallace, a fellow naturalist who was at that time in what is now Indonesia. Unlike Darwin, Wallace had no inheritance and made his living collecting biological specimens all over the tropics. Wallace, like Darwin before him, had read Thomas Malthus' speculation about population growth: how a biological population will tend to grow faster than the available food supply. Also, like Darwin, he saw that limits on resources must lead to competition between individuals and "survival of the fittest."[19] Wallace's letter contained the draft of a paper that he asked Darwin to pass on for publication if he thought it any good. The paper was titled "On the tendency of varieties to depart indefinitely from the original type." It even used Darwin's own phrase, *natural selection,* to describe the very same process that Darwin had discovered, and been gathering data to support, for some twenty years. In a letter to his friend Lyell, Darwin wrote: "I never saw a more striking coincidence. If Wallace had my M.S. sketch written out in 1842 he could not have made a better short abstract! Even his terms now stand as Heads of my Chapters."

Darwin was in a fix. He had published nothing on his idea. By modern standards, he should have ceded priority to Wallace. On the other hand, by more responsible standards, Darwin deserved credit for his years of meticulous research and unwillingness to publish until his point could be proved. After consulting with his friends, Darwin wrote a short paper that was given in 1858 at the Linnaean Society meeting at the same time as Wallace's. Neither attended and the papers were presented by Darwin's friends Charles Lyell, author of the influential *Principles of Geology*[20] and Joseph Hooker, at that time working on botanical specimens, including Darwin's *Beagle* trove. In a classic 'miss,' the Society's president, at the end of the year, commented: "The year which has passed has not, indeed, been marked by any of those striking discoveries which at once revolutionize, so to speak, the department of science on which they bear." (Wallace was very happy with Darwin's handling of his paper, entitling one of his subsequent books *Darwinism*.)

Darwin immediately went to work and produced in 1859 an 'abstract'—*The Origin of Species*—of what he had planned as a much longer book. The shortened (but still 600+ pages!) *Origin* was a best-seller.

Could this story be repeated today? Unlikely. How many scientists now are able to work as long and as carefully as Darwin before professional imperatives drive them prematurely into print? How few have the security and restraint to wait to publish until they have really proved something substantial?

In an era where science has gone from being the vocation of a tiny minority to a career for the many, the conventional journal system with its anonymous peer review, long publication delays and potential conflicts of interest, is beginning to show its limitations.[21] It will likely be superseded by some form of open-source rapid publication.[22] Perhaps that will mitigate a pressure to publish that seems to be the source of many problems with contemporary social science.

The Tea-Party Study

Unfortunately, even well-intentioned research using the NHST method can mislead. Here is a typical example, from a study reported on National Public Radio in 2016: *Threats to Racial Status Promote Tea Party Support among White Americans*.[23] The paper comprises five experiments; I will just discuss Experiments 1 and 4.

The first experiment is the simplest to understand. It had a total of 356 subjects, 255 of whom were white; they were divided randomly into two groups. The key manipulation involved pictures. All subjects were shown three pictures: two celebrities (William Shatner, Jay Leno) and a modified version of an official picture of President Obama. "To make more or less salient his African-American heritage, participants were randomly assigned to see a version of the picture in which Obama's skin was either artificially lightened or darkened." So, half the subjects saw the lightened picture, half the darkened one. After viewing the pictures, all subjects were asked a series of yes-no questions, one of which was: "do you consider yourself a supporter of the Tea Party?" The null hypothesis is that using the Obama-picture version the subject saw made no difference to his/her response to this question.

The conclusion which caught the attention of NPR was this: "white participants in the Dark Obama Prime condition were significantly more likely to report that they supported the Tea Party (22% [= 28 people] than white participants assigned to the Light Obama Prime condition (12% [= 15 people]; $\chi2$ (1) = 4.73, φ = .139, p = .03 [see table, totals rounded]). This result supports our prediction that white Americans would be more likely to support the Tea Party if Barack Obama's African-American heritage was made salient to them." (Male–female differences and the reaction of the nonwhite participants were not reported.)

		Tea Party fans?
total white subjects	256	
shown dark Obama pic	128	**28**
shown light Obama pic	128	**15**

A fourth study addressed the same issue as the first one by presenting avowed Tea-Party supporters with the same demographic data presented in two ways. One graph appeared to show a slow decline from 2000 to 2020 (estimated) in the proportion of whites in the US population; the other was an extended and exaggerated version of the same data, extrapolated to 2060, which appeared to show a large decline in the proportion of whites. They found greater allegiance to the Tea Party, assessed via questionnaire, for the group shown the exaggerated graph.

Race is a politically sensitive subject in the United States. I use this example just for that reason, to show how influential these experiments may be and thus how important it is to understand their limitations.

There are several obvious questions to ask about a NHST study like this: (1) How reliable is the significance measure? (2) Could the study be replicated? (3) Assuming that (1) and (2) can be answered satisfactorily, do the authors' conclusions follow?

Models and Significance Level

The results of Experiment 1 were significant at the 3% level, according to the popular Chi-square ($\chi 2$) statistical test, representing a difference of 13 more people (out of around 128 in the group) who answered "yes" to the Tea-Party question in Dark Obama group (table). In other words, if as few as six or eight people out of the 255 in the study, had changed their minds, the difference observed would not have been statistically significant.

To measure the significance level, the authors need to know the sample space, the range of variation of responses to their yes-no questionnaire in the absence of any manipulation—after seeing just the light-Obama picture, for example. This is no easy matter. The least-flawed method, the only method which would yield a sample space with the same precision as Fisher's tea-lady experiment, is as follows. Do a large 'control' experiment in which the very same questionnaire is administered repeatedly to different groups of 250 or so white people exposed just to the light-Obama picture—100 such groups would mean 25,000 subjects, a very large experiment indeed. The number of each group of 250 responding "yes" (T) to the Tea-Party tick box would be counted, and the numbers grouped in categories (between 1–10, 10–20, …, 24–250) and plotted. The result will presumably (although even this is not certain) be something like the typical bell-shaped 'normal' distribution. Most of the groups will be clustered around the

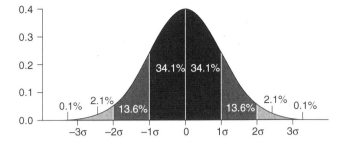

The normal distribution, showing one, two and three standard deviations from the mean. Percentages show the % of sample in the indicated regions. X-axis shows scores in categories above and below the mean set at zero.

Source: Wikipedia, https://en.wikipedia.org/wiki/Standard_deviation

mean, the average, for the entire population of 25,000, with a declining number in the outer categories.

This is necessarily just a thought experiment, since it is too large to be practicable—in this case and many others like it. Exact specification of the sample space is usually impossible. Hence, there is no alternative to testing the null hypothesis with the aid of what is in effect another hypothesis: a *model* of the sample space. A multitude of statistical models are available. The methods that incorporate them tend to be used as tools, tautologies like arithmetic or algebra, as if they were essentially neutral about the outcome. But, as we will see, they are just as hypothetical, just as uncertain, just as biased, as the hypothesis being tested.

What will be the mean of the distribution we might have obtained with our thought experiment? Our best guess is that it will be close to the experimental result for the light-Obama group: 15. Our guess here is as likely to be true as a guess in the tea experiment that 50-50 chance of T or M will yield exactly four of each out of eight cups. In other words, we really don't know exactly what the experimental mean should be.

How far out on such a distribution is the actual result for the dark-Obama group (28)? To reach the 5% significance level it would need to be more than about two standard deviations from the mean, but of course we don't know the distribution so cannot answer the question for sure.

This conundrum illustrates the problem for essentially all the commonly used statistical procedures: They *don't know the sample space*. In the absence of any direct measurement, all depends on an underlying model.

Let's look at one possible model which captures as simply as possible the idea that subjects are *not* influenced by the picture they see, i.e., the null hypothesis. Suppose that each individual chooses independently and makes his/her choice as if tossing a coin. Since the two options are not equiprobable—most subjects are not Tea-party fans—the coins must be biased: the T-P probability must be less than 50%.

We don't know the bias of the coin but we can pose the question in this way: is there a biased coin that could plausibly give us results as extreme as the results actually obtained in the experiment, namely 15 (light-Obama) and 28 (dark-Obama) T-P choices out of 255 tosses?

It is easy to answer this question by checking with a *binomial-distribution* calculator. The binomial distribution describes the relative frequency of all the different outcomes when you toss a coin a certain number of times. For three tosses of an unbiased ($p = 0.5$) coin, for example, the eight possible outcomes are:

Outcome	Frequency of H
HHH	3
HHT	2
HTH	2
HTT	1
TTT	0
TTH	1
THT	1
THH	2

1 3 3 1
0 1 2 3
Overall frequency of H

The probability of just one H is 3/8, of two H, 3/8, and of three or zero H, 1/8. The binomial theorem generates a similar distribution for any number of coin tosses, all symmetrical around a peak at p.

How does this help us with the original question? We have 255 hypothetical tosses of a biased coin. For a given bias, the binomial theorem can tell us the probability of a number of T-P choices greater or less than a given number. What should p, the bias, be? We don't know.

So, to repeat the question: is there a bias value which would yield the observed T-P choice numbers, both high and low, with a relatively high probability? The answer is 'yes.' A bias of $p = 0.085$ 'T-P' versus 0.915 'not-T-P' yields results ≤ 15 with probability 0.07 and ≥ 28 with probability 0.098. Both of these numbers are above the arbitrary 'not-significant' 0.05 threshold. In other words, it is quite possible that the two Obama pictures had no effect on subjects' preference for the Tea-Party, even by the generous 5% Fisher standard.

It is very easy in statistics-world to lose sight of the individual subject and what he or she might plausibly be doing. Yet the group data depend entirely on individual decisions. Any model should incorporate what little we know about the individuals who make up the group. So, let us look again at our model, which assumes that every subject is not just biased but biased to exactly the same extent: 0.085. It would surely be more realistic to assume some variation in individual bias, in which case, the spread of the distribution would certainly be greater than for the fixed-bias case—and extreme results thus even more likely. So, given the most plausible model, the results are almost certainly not significant, even at the 5% level.

It should be obvious by now that it is quite possible to 'game' the system, whether intentionally or not, to find a model that makes your data cross the 'publishable' threshold. The incentives to do this are often very strong.

There are other models for data like this. Most popular is the Chi-square test, invented by Karl Pearson, which is what the authors used go get their 3% significance. A better test is the binomial-based Fisher's exact test, because it *is* exact. It shows the results to be significant, but at the 5%, not 3%, level. But again, like my analysis, Fisher assumes the same binomial bias for each individual so, again, if each individual has a different bias, the result would probably not be significant, even at the 5% level. The very best test would be, as usual, replication. Otherwise, the sample space can only be estimated—guessed—not proved.

The multitude of methods, even for this relatively simple problem, shows just how uncertain these methods are. There is just one method for measuring temperature or weight or force or many other scientific quantities (or, to say the same thing differently, different methods all give the same answer). Statistics is rarely like that, and studies that must use statistics are a very long way from Dr. Snow and the pump handle or Joseph Priestley and the discovery of oxygen.

Is the Tea-Party experiment replicable? Experiment 4, using a different procedure, is an attempt to get at the core idea—that Tea Partiers are motived to some extent by racial issues. But exact replication is unlikely, if only because a new set of subjects may now understand the purpose of the study and might modify their responses accordingly. Problems with the whole NHST method, (remember the 1000 experiments on ineffective drugs) also suggest that any attempt at exact replication is likely to fail. In any case, literal replications are rarely done, because they are hard to get published (although attempts are being made[24] to remedy this).

Strength and Elasticity of Attitudes

To evaluate the social relevance of research, like the Tea-Party/Obama study, that bears on important issues, we need to know at least two things. First, just how important is, say, a distaste for margarine versus butter, or a preference for African Americans over whites, to individual X? Does X really care or is he just ticking a box to be polite? Archie Bunker may show a tick-box bias against African Americans, yet he may be perfectly willing to vote for one or even have his daughter marry one. Beliefs have a *strength* which is measured only imperfectly by surveys.

But beliefs also have a second property I will call *elasticity*. Elasticity is a familiar concept to economists. The simple law of supply and demand dictates that a small increase in price will (usually) cause a small decrease in demand (i.e., number bought). Elasticity is just the ratio of the small change in demand to the small change in: price, both expressed as ratios:

$$E = \frac{\Delta Q / Q}{\Delta P / P}$$

Where Q is the quantity of a good and P is its price. If price elasticity is great, then a small change in price will cause a large change in quantity bought, in demand. The equation has no direct application to strength of belief. But the idea that a variable, whether belief or demand, has not just a strength but a susceptibility to change, does apply.

It is a principle more or less universally acknowledged that the more you have of something the less will you desire still more of it. Demand depends on the value, the *utility*, of a good. As I will discuss at more length in Chapter 6, marginal utility—the value of each little ΔQ—usually declines with value of Q: large at first (one slice of pizza) and decreasing as the amount consumed increases (the 10th slice). So, price elasticity will be small if the stock of the good is small but larger as the stock increases.

Much the same seems is true of *beliefs*. There seems to be a parallel between amount of a good and amount of knowledge about something; and also, a parallel between degree of belief and price. If you own very little of something desirable, you will be willing to pay a lot for a small increment: marginal utility and price elasticity, will be high Conversely, if you already own a lot, an increment will be much less attractive and you will not want to pay much for it: price elasticity will be low.

Similarly, if you know very little about something, then a small addition to your stock may cause a large change in your belief. Conversely, if you already know a lot, a small increment in your knowledge is likely to have little effect. In other words, as Reverend Thomas Bayes pointed out a century and a half ago in connection with probability, the effect of new information depends on the amount of information you already have. If you know a lot, a little more information won't change much. Conversely, a little knowledge is easily augmented. So, a bigot is not a bigot is not a bigot. A person whose beliefs about, say Africans, have formed through hearsay is likely to be much more impressed by personal experience with an actual African than someone who has acquired his beliefs through a rich personal history. As a tendentious witticism puts it: "A conservative is a liberal who has been mugged by experience."

Thus, we should not be too surprised at a recent report[25] that more information, whether positive or negative, reduces racial discrimination:

> Recent research has found widespread discrimination by [Airbnb] hosts against guests of certain races in online marketplaces … We conducted two randomized field experiments among 1,256 hosts on Airbnb by creating fictitious guest accounts and sending accommodation requests to them … requests from guests with distinctively African American names are 19 percentage points less likely to be accepted than those with distinctively white names. However, a public review [whether positive or negative] posted on a guest's page mitigates discrimination: … the acceptance rates of guest accounts with distinctively white and African American names are

statistically indistinguishable. Our finding is consistent with statistical discrimination: when lacking perfect information, hosts infer the quality of a guest by race and make rental decisions based on the average predicted quality of each racial group; when enough information is shared, hosts do not need to infer guests' quality from their race, and discrimination is eliminated.

In other words, the modest racial bias Airbnb landlords showed at first was very elastic; it could be dissipated by almost any credible information, whether positive or negative. Unfortunately, measuring elasticity may be difficult because people may not be aware of how susceptible to change their beliefs are. They only know when their beliefs are actually changed by new information.

What Can We Conclude about the Tea-Party Experiment?

The authors' overall conclusion is that "these findings are consistent with a view of popular support for the Tea Party as resulting, in part, from threats to the status of whites in America." Does it follow from their results? Well, the data are certainly 'consistent with' their hypothesis. But of course, the important questions are quantitative: just *how important* is 'racial threat' (also termed 'racial status anxiety,' neither clearly defined) to the Tea-Party position; how strong is the bias? Ticking a box in a questionnaire is not exactly a commitment. In other words, we need to know if not the whole story at least something close to it, to evaluate the importance of race to white people's political decisions. If President Obama were miraculously revealed to be really a white guy, most Tea Party fans are unlikely to give up their allegiance. The racial component of the Tea Party may actually be minimal, even if the results of this experiment were indisputably both significant and repeatable.

The second question is about elasticity: how changeable, how responsive to new evidence, is the anti-Obama bias? Would these people change their view if given new information? We don't know the answer to this question either.

In short, the Tea-Party experiment, like many similar studies, is no more than an *uninterpretable fragment*. The effect may not be real and even if it is, the amount that racial bias might contribute to Tea-Partiers allegiance is unknown. Nevertheless, the study made it on to national media—which is a real problem, because the results are presented not as opinion or as a hypothesis, but as *science*. Is racial bias at the heart of the Tea Party? The apparent "yes" conclusion of this experiment will inevitably have too much influence on how people think about these rather fraught issues.

Whither NHST?

Even if the probabilities yielded by standard statistical methods are accurate, there is still a serious problem with the NHST method. In the 2005 article entitled

"Why Most Published Research Findings Are False" that I referred to earlier, researcher John Ioannidis, concluded: "simulations show that for most study designs and settings, it is more likely for a research claim to be false than true. Moreover, for many current scientific fields, claimed research findings may often be simply accurate measures of the prevailing bias." (The findings of the Tea-Party experiment confirmed the views of influential critics, who thought the Tea-Party racist.)

The failures of the NHST method have shown up most strikingly in drug studies, efforts to find medications to cure disease. A 2011 *Wall Street Journal* article[26] concluded "This is one of medicine's dirty secrets: Most results, including those that appear in top-flight peer-reviewed journals, can't be reproduced. In September, Bayer published a study describing how it had halted nearly two-thirds of its early drug target projects because in-house experiments failed to match claims made in the literature."

In retrospect, these failures could have been predicted. Drug companies and academic research labs test thousands of potentially curative chemical compounds. As I pointed out earlier, if 1000 ineffective drugs are tested, then by chance, about 50 studies will surpass the generous 5% significance level. If some of these tests were done in academic labs, they will be published. Negative results, tests that show ineffective drugs to be ineffective, are unlikely to be published (again, efforts to change that bias are being made). Since the drugs in all these hypothetical studies are in fact ineffective, attempts to replicate their effects are likely to fail, as the *Journal* reported.

The authors of any individual study *will not know about the whole sample space*. They will not know how many drugs (similar drugs? All drugs?) have been tried nor how many tests have failed. They will not know the *prior probability* of success. They will not, therefore, be in a position to assess the likelihood that their own 'statistically significant' one-shot attempt is likely to be unreplicable.

Can this problem be fixed? Many technical discussions were prompted by Ioannidis' article and the matter is still in flux. One remedy is to follow physics and set the publication threshold very much lower than 5%.[27] If single (i.e., unreplicated) studies with NHST probabilities greater than, for example, 0.005 (rather than 0.05) were deemed unpublishable, the social science and pharmacology literatures would shrink to manageable proportions and their ratio of signal to noise would certainly rise. This solution is not popular because it would seriously impair researchers' ability to get published. The 5% standard does after all guarantee that effort, if not scientific progress, will not go unrewarded.

But a better solution would be to abandon the whole idea that a single number, be it significance level or even effect size, is sufficient to evaluate the 'truth' of a research result. A study should be evaluated as a contribution to the advancement of knowledge. This is very difficult to do, of course; and not something that can be easily measured. A study must be evaluated in relation to the rest of the field, or fields, on which it touches: what does it add to the existing sum of knowledge? What I have called uninterpretable fragments, like the Tea-Party study, should never see the light of day, although they might form part of a group of studies that arrive at a meaningful conclusion about how beliefs are formed and the role they play in political action.

Applied—policy related—studies should be evaluated differently than basic research. A big and significant causal effect might be a legitimate guide to, say, health policy. A big and significant noncausal effect should have no effect on policy but might provide hypotheses to guide additional experiments. Or it may add a piece to a jig-saw puzzle—Bacon's *alphabet* again—of other correlations that together paint a convincing picture. A small effect, significant or not, should not be acceptable as an isolated contribution but might serve as a guide for further basic research. It might be an aid to making a forced decision, as in the Fisher example. The point is that applied research is aimed at a defined *goal*, building a rocket, constructing a fusion reactor, or curing a disease. It's easy to know whether a given research program has succeeded or not. Did the rocket fly? Did the reactor work? In basic research the aim is *understanding*, something much harder to define in advance.

Perhaps we should look again at the history of social and psychological science as a guide to what is worth publishing and what is not—at what is real knowledge as opposed to what is not.

Notes

1 "The scientist must find order; science is made with facts as a house is made with stones, but a pile of facts is not a science as a pile of stones is not a house."

2 Skinner, B. F. (1956) A case history in scientific method. *American Psychologist*, 11(5), pp. 221–233; Grindley, G. C. (1923) The formation of a simple habit in guinea pigs. *British Journal of Psychology*, 23(2), pp. 127–147.

3 Markman, S., Leitner, S., Catchpole, C., Barnsley, S., Müller, C. T., Pascoe, D., and Buchanan, K. L. (2008) Pollutants increase song complexity and the volume of the brain area HVC in a songbird. *PLoS One* 3(2): e1674. doi:10.1371/journal.pone.0001674

4 https://physics.aps.org/featured-article-pdf/10.1103/PhysRevLett.116.061102

5 https://academic.oup.com/ije/article/32/6/938/775148/RA-Fisher-statistician-and-geneticist

6 I use true/false in this discussion for simplicity and because it is conventional. But bear in mind that in experimental science *true* just means *replicable*.

7 Epidemiologists may also participate in experimental tests of observed correlations, when such tests are possible. Often, unfortunately, tests are not possible, for the reasons I describe.

8 www.hsph.harvard.edu/news/press-releases/plant-based-diet-reduced-diabetes-risk-hu-satija/

9 Ioannidis, J. P. A. (2005) Why most published research findings are false. *PLoS Medicine*, 2(8) Ioannidis, Cohen, 1990. See also www.theatlantic.com/science/archive/2015/08/psychology-studies-reliability-reproducability-nosek/402466/; www.theatlantic.com/science/archive/2016/03/psychologys-replication-crisis-cant-be-wished-away/472272/

10 http://thelancet.com/journals/lancet/article/PIIS0140-6736(16)32399-6/fulltext

11   I emphasize the number of authors because multiple authorship has much increased in recent years. One reason? It gives more people a publication to add to their CV: See "Bad incentives, bad results" below.

12   Staddon, John (2013/2014 US) *Unlucky strike: Private health and the science, law, and politics of smoking.* Buckingham, UK: University of Buckingham Press.

13   Smaldino P. E., and McElreath R. (2016) The natural selection of bad science. *Royal Society Open Science* 3(9), p. 160384. http://dx.doi.org/10.1098/rsos.160384. See also Deep impact: Unintended consequences of journal rank. Björn Brembs, Katherine Button, and Marcus Munafò. Available at: https://doi.org/10.3389/fnhum.2013.00291

14   Buranyi, S. (2017). The High-Tech War on Science Fraud. *The Guardian.* Available at: www.theguardian.com/science/2017/feb/01/high-tech-war-on-science

15   Spoiled science. *The Chronicle of Higher Education,* 3/18/2017. See also Prof. Wansink's response: http://foodpsychology.cornell.edu/note-brian-wansink-research#feb. The strategy discussed here is called *p-hacking.* For example, imagine a multi-variable experiment in which standard NHST comparison shows no significant ($p < 0.05$) difference for variable A between experimental and control conditions: the authors' hypothesis is not supported. But if other measures, B, C, D, etc. have also been taken, possibly a significant difference can be found for one of these? Such a result might be useful in designing a future direct test. But the study can also be 'salvaged' by inventing a hypothesis compatible with this result and re-writing the paper as a test of this new hypothesis. *P-hacking* is not legitimate, scientifically or ethically. https://en.wikipedia.org/wiki/Data_dredging#/media/File:Spurious_correlations_-_spelling_bee_spiders.svg

16   "Least Publishable Unit".

17   www.newyorker.com/tech/elements/paging-dr-fraud-the-fake-publishers-that-are-ruining-science;www.nature.com/news/predatory-journals-recruit-fake-editor-1.216 62?WT.mc_id=SFB_NNEWS_1508_RHBox

18   Here is one from a few weeks ago: *Alzheimer's and Parkinson's Diseases,* preceded a week or so earlier by *Journal of Phylogenetics & Evolutionary Biology.* A recent and especially seductive example is the following:

> Dear Dr. Staddon, One of the articles you authored a while ago caught my attention and I am hoping to discuss with you publishing a follow-up article, or even a review article in the Internal Medicne [sic] Review. The article was entitled 'The dynamics of successive induction in larval zebrafish.' I am sure our readers would find an article that continues this work valuable. The parameters of the article are flexible and I am happy to help in any way I can … Lisseth Tovar, M. D., Senior Editor, Internal Medicine Review.

I am still puzzling over the links that I may have missed between internal medicine and the reflex behavior of zebrafish larvae. But then just this week, I got an even more exciting offer: to be an actual editor for the *Journal of Plant Physiology* (my emphasis).

19   19th century philosopher and popular science and politics writer Herbert Spencer's phrase, not Darwin's.

20   Lyell's *Principles* changed the *zeitgeist* of geology, previously dominated by *catastrophism,* the idea that natural features are the product of disasters like great floods and volcanic upheavals. Lyell emphasized *uniformitarianism,* the idea that the earth's features can be explained by processes going on today, many slow, like erosion and changes in the biosphere. Uniformitarianism much influenced Darwin's thinking about evolution.

21   www.economist.com/news/science-and-technology/21719438-about-change-\findings-medical-research-are-disseminated-too

22   www.theguardian.com/commentisfree/2013/dec/09/how-journals-nature-science-cell-damage-science

23   www.npr.org/2016/07/14/486063689/study-explores-links-between-politics-and-racial-bias

24 http://science.sciencemag.org/content/349/6251/aac4716.full?ijkey=1xgFoCnpLsw pk&keytype=ref&siteid=sci

25 Discrimination with incomplete information in the sharing economy: Field evidence from Airbnb. Ruomeng Cui, Jun Li, and Dennis J. Zhang. Working paper: https://papers.ssrn.com/sol3/papers.cfm?abstract_id=2882982

26 Scientists' elusive goal: Reproducing study results: www.wsj.com/articles/SB10001424052970203764804577059841672541590

27 There are influential moves to increase the significance level that is acceptable for publication (e.g., from p = 0.05 to p = 0.005 or even lower): www.nature.com/news/big-names-in-statistics-want-to-shake-up-much-maligned-p-value-1.22375?WT.mc_id=FBK_NA_1707_FHNEWSPVALUE_PORTFOLIO

4

SOCIAL SCIENCE

Psychology

The difficulty of testing hypotheses in the social sciences has led to an abbreviation of the scientific method in which [experimental test] is simply omitted. Plausible hypotheses are merely set down as facts without further ado. To a certain deplorable extent this same practice occurs in medicine as well.

E. Bright Wilson

How do they know?

Anon

Is 'social science' an oxymoron? One hopes not, but the temptation to see it as a contradiction in terms is quite natural. Indeed, many philosophers claim social science is intrinsically different from 'natural' science and should be judged by different standards.[1] It follows, and should follow, they say, different—more relaxed—rules of investigation and proof. This idea has been advanced most persuasively in economics, about which more later. But there is no need to give social science an 'out' like this. It just suffers from three intractable problems not shared to anything like the same extent by physical and even biological science. These difficulties can alone account for the very many differences—of precision, of testability, of validity—between social and natural science. The differences arise from difficulties that are practical rather than conceptual. The problems are easy to identify, though hard to overcome. But they do not justify giving up the ghost and settling for weaker standards in social science.

Three Problems for Social Science

Ethical

Experimental investigation is very difficult. Doing experiments with people on topics about which they care, such as health, social arrangements, and money, always raises ethical problems. Experiments on choice require that people be subjected to consequences—reward and/or punishment. These consequences must be either notional (for example, points, gold stars) or trivial if they are not to raise serious ethical issues. Strong 'reinforcers' (not to mention punishers!) are either impossibly expensive or ethically problematic. Developmental psychology—experiments on child-rearing practices, for example—encounters similar problems. The kind of control over subject matter that is routine in chemistry and physics is simply impossible in most of social science.

Causal

Because meaningful experiments are usually impossible, social science relies heavily on correlational studies. But as I have already pointed out, possibly *ad nauseam*, surveys, epidemiology, and the like, cannot prove causation. Only if many independent information sources converge in a compelling way can we have much confidence in identifying social causes in the way causes can be identified in, say, physics or biology. Very little social psychology involves broad-ranging studies of this sort. Instead much of the literature consists of papers with a handful of experiments under highly artificial conditions. From these fragments, ill-founded conclusions are often drawn about important social and political issues.

Time scale

Human beings and human societies—complex systems in general—have memories. The effects of any environmental or social change will often take many years to play out. I already discussed the problem that time delays pose for a relatively simple problem: the presumed cancer-causing effects of smoking. Effects are usually delayed by several decades and proper experiments cannot be done for ethical reasons. The same is true of, for example, the effects of changes in child-rearing, in educational curricula, or in town planning and transportation policy. Social-science experiments over the time span needed to assess the effects of changes like this are essentially nonexistent. Even correlational studies are rare. Usually denied rigorous methods of investigation, social scientists have resorted to less satisfactory methods.

A method they might have used, but have largely neglected, is orderly systematic observation. Pioneer biologists, like Buffon, Linnaeus, Haeckel, Cuvier, and even Darwin in his early years, were primarily collectors and systematizers. They collected plants and animals and looked at their structure, reproductive habits, and lifestyle. They tried to arrange specimens and other data in sensible ways. As patterns began to emerge, laws were proposed. Some could be tested. Others, like evolution by natural selection, emerged as organizing principles.

The method of induction, based on systematic observation, has been too little used by social psychologists. I recently heard of an unpublished observational study on social interaction which catalogued the strategies used by each party.[2] The two groups were European sailors on noncolonial mapping and surveying missions in the 18th and 19th centuries, and various Pacific islanders. The behavior of sailors and islanders could be assessed from the careful logs kept by ships' captains. Interesting patterns emerged.

A published example of induction in social science is the work of Swiss developmental psychologist, Jean Piaget, who began as a biologist. He made his mark by studying carefully a relatively small number of children through their early years. He did few experiments and no experiments that used group-average data. When he worked at a school, he graded early intelligence tests and noticed that there was a pattern to the errors that children made. From this he concluded that there was a structure to mental development with human capacities developing in a fixed order.

Piaget's method of direct observation is not always practical. For example, the life-long effects of different kinds of child-rearing cannot easily be studied directly. Researchers have neither the time nor the opportunity to watch individuals throughout their lives. There is an alternative—less satisfactory than direct observation but more practical. A student of child-rearing might first of all look through a wide range of biographies. How were these people raised? How did they develop as adults? Are there patterns—do certain kinds of early history tend to go along with certain life choices? The method is imperfect to be sure. The sample will be too small and far from random: nonentities rarely attract biographers. But the results might serve to rule out some things and suggest others. Large-scale longitudinal surveys are another option although they are still correlational and can tell us little about individual psychology.[3]

Observational methods like this—sometimes disorganized, always correlational, following no accredited 'gold standard'—are rarely used in social psychology. Instead, the prestige of the experimental method and its legitimation by sources of science funding, has pushed it to the forefront. Given the ethical, causal, and time scale difficulties I mentioned earlier, the most popular option is short-term experiments involving weak manipulations, linked by some kind of theory to presumed long-term effects. Speculation—theory—either in a verbal, non-quantitative form, or what might be called *a prioristic,* often quantitative, theory is another option. The first tactic is common in social psychology; the second in economics. I begin with social psychology.

Social and Personality Psychology

Science often advances by inventing imaginary entities. Genes and atoms, neutrons and quarks, were all at first hypothetical, proposed before they could be measured directly. But each one had properties that implied observable effects. A similar strategy has been followed in much of social and personality psychology. One difference is that the imagined entities often derive from what has been called 'folk psychology,' the common-sense language that human beings use to communicate ideas about their own behavior/ideas/attitudes, or the behavior of others, to one another. We will see an example of this strategy in the great Adam Smith's *Theory of the moral sentiments* (1759), which I discuss at more length in the next chapter. Smith used terms like *sympathy* (empathy), *sorrow, self-love, disdain, disgust, respect*, and the *sense of propriety*. Modern social psychologists favor terms that sound more scientific, like *self-efficacy, cognitive dissonance*, or *the fundamental attribution error*. The question for all these constructs is the same: what do they imply for the third-party world? What are their measurable effects? What do they add to folk psychology and common sense? Above all, how can they be tested?

Intelligence

> *It is not the strongest of the species that survive, nor the most intelligent, but the one most responsive to change.*[4]

Personality measurement has borrowed extensively from folk psychology, especially in the study of intellectual ability. One of the earliest scientifically inclined psychological inventions is the *intelligence quotient* (IQ). *Intelligence* is not a technical term. People have thought of each other as 'smart' or 'dumb' as long humans have existed. Intelligence was eventually 'operationalized,' in the jargon of the time,[5] as answers to a set of questions: the IQ test. The questions had to be derived intuitively, as plausible measures of how smart the exam-taker is. There is no rule that can tell us ahead of time what will be a valid question to put on an IQ test. They are just invented.

How do the inventors know that they have got the questions right? IQ tests are usually evaluated by two criteria: (1) Is it just another exam? In other words, can a student study for it and do much better with more practice? If so, the test is presumably more a measure of accomplishment than of ability. If students cannot 'learn the test' then it *may* be a good IQ test. (2) We can then ask: does it do the job we wanted it to do? If the test is supposed to measure 'intelligence,' then does it predict the things we associate with intelligence (folk psychology again!), like, for example, academic achievement or success in certain kinds of job? The prediction will never be perfect, of course, because success in any field depends on effort and luck as well as ability. IQ tests do not measure motivation.

People can study for IQ tests and study does help a bit, although much less than for the usual subject-based exam. And there are completely nonverbal IQ tests, like Raven's Progressive Matrices, for which study is less help. So apparently the tests are measuring something relatively fixed, some ability—call it intelligence if you like.

The tests are designed such that the older you get—up to about 16 or so—the more questions you are likely to get right. Intelligence quotient is then 'normed' to your age: a ten-year-old who answers as many questions as the average

13-year-old is then said to have a *mental age* of 13 and an IQ of 13/10 x 100 = 130.

Existing IQ tests do seem to predict things like college grades. I say 'seem' because to adequately test the idea, students would have to be admitted to a college at random or at least in ways that took little account of IQ. That may have been more or less true at one time, when your social origins—wealth, connections—were more important than your brain for admission to an elite college. But nowadays most applicants to are ranked according to IQ-like criteria like the SAT (Scholastic Aptitude Test) and similar tests. Which means the IQ range of those actually admitted is probably limited—which will tend to abolish the correlation between academic results and IQ. To illustrate the problem: Height is an asset to a basketball player. Most professional basketball players are tall. But there is no correlation between height and game performance in that population, because the range of heights is limited: all, or almost all, professional players are very tall. The range of IQs in competitive colleges is also limited: most students are high-IQ. Given all the other factors that affect grade—effort, type of course, special aptitudes, etc.—little correlation of IQ with grade would be expected within populations in most elite colleges—and little is found.[6] But that most emphatically does not mean that IQ is unrelated to academic performance.

The original IQ test was invented by Frenchman Alfred Binet in 1904. His test had an ameliorative purpose: to identify children—pupils—with *low* intelligence, who could not be expected to handle a regular school and needed remedial treatment. Now, of course, IQ tests are often used for an opposite reason: to identify *high* talent, a function for which they are probably less well suited. Either way, IQ research is a form of technology, intended to solve a practical problem, rather than basic science, intended to understand the workings of brain and mind.

IQ tests do work pretty well as predictors and have been used, in one form or another, as selection tools for many years. Any meritocratic filter encounters political problems, the most recent example being a UK move to bring back "grammar schools," which stream exam-bright kids into a college-bound track. IQ tests also regularly excite controversy. Nevertheless, in one form or another, they are likely to be with us for some time to come.

Research on intelligence began well before the invention of the first test, with polymath Victorian genius, Francis Galton (1822–1911), Charles Darwin's young half-cousin. Galton made major contributions to statistics (correlation, regression to the mean), the classification of fingerprints and meteorology, among other things. He explained his interest in heredity and exceptional talent as follows:

> THE idea of investigating the subject of hereditary genius occurred to me during the course of a purely ethnological inquiry, into the mental peculiarities of different races; when the fact, that characteristics cling to families, was so frequently forced on my notice as to induce me to pay especial attention to that branch of the subject. I began by thinking over the

dispositions and achievements of my contemporaries at school, at college, and in after life, and was surprised to find how frequently ability seemed to go by descent.[7]

Galton just looked at the careers of his subjects. He devised no mental test—although he looked to see if some objective characteristic, like reaction time, was associated with intellectual ability (it is, a little). Nor did he know much about the mechanisms of heredity—genes only came on the scene much later. He did know that environment and heredity are usually mixed: he coined the phrase nature versus nurture. People in the same family share both their heredity and their environment. So, no strong hereditarian argument can be made by observing natural families. Nevertheless, Galton introduced two themes—heredity and race differences—that were to form part of, and bedevil, research on human intelligence ever afterwards:

> I PROPOSE to show in this book that a man's natural abilities are derived by inheritance, under exactly the same limitations as are the form and physical features of the whole organic world. Consequently, as it is easy, notwithstanding those limitations, to obtain by careful selection a permanent breed of dogs or horses gifted with peculiar powers of running, or of doing anything else, so it would be quite practicable to produce a highly-gifted race of men by judicious marriages during several consecutive generations.

From this innocent beginning arose a keen interest in the intelligence of 'primitive' races, a Department of Eugenics in University College, London (founded 1933, abolished as a separate entity by 1965), and seemingly endless conflict over the relative importance of genetic differences versus the legacy of slavery and tribal culture in the persistent black-white IQ differences in the United States.

Once the idea of intelligence as a researchable subject was accepted, arguments began about what it is. Is it unitary—the *'g'* factor—or are there many intelligences? The multiple-intelligence theme was strongly advocated by one eminent researcher who admitted he had performed poorly on IQ tests as a child. He proposed that there is not just one intelligence but three. Others proposed still more 'intelligences': social intelligence, emotional intelligence, and a few others; and there is yet another distinction between *fluid* and *crystallized* intelligence. Black icon, boxer Mohammed Ali, shows how hard it is to define intelligence. He reportedly had a low IQ, but was obviously talented in many ways, mental as well as physical, that the test does not measure. Artificial intelligence is another illustration. Recently 'deep learning' AI researchers have begun to produce programs that perform better than humans on IQ tests[8] (deep learning is the technology that enabled IBM's *Watson* to beat human *Jeopardy* champions in 2011). It is unlikely that the computer operates in exactly the same way as the human being. Nor is it likely to be able to do things that are routine for intelligent people—write an essay or compose a poem, for example—although with additional programming, such things may also be within its reach. The point is that the ability to do well on IQ tests does not automatically entail an ability to do other 'intelligent' things, because there are probably many computational ways to do well on IQ tests. It may not be possible to gain much understanding of human intelligence from IQ measures alone.

Research on intelligence has become needlessly controversial. To explore it would take us well beyond the 'method' theme of this little book. But it is perhaps worth looking at two contentious issues. Just what would it take to settle the vexed question of the heritability of IQ and the reality or otherwise of race differences in intelligence?

What is 'intelligence'? Is it, as one eminent psychologist asserted, "just what the tests measure"? Well, yes, it is in fact, because there is no other generally agreed definition. Intelligence, for the purpose of this discussion, *is* what we measure with suitably validated tests. The 19th century physicist Lord Kelvin famously said something along the following lines: if you can't measure it, you don't understand it. His aphorism may be true, but its converse is not. It is not true that if you *can* measure it, you *do* understand it. Measurement does not equal knowledge. The fact that we can get a score on a test and call it 'intelligence' does not mean that we understand intelligence.

With these caveats, let's look at nature versus nurture: is intelligence inherited, as Galton proposed, or is it entirely a product of upbringing, as the influential 17th century Enlightenment thinker, John Locke (see *tabula rasa*)—and many 20th century behaviorists—believed? Now, we know a bit about genetics, the word *heritability* has acquired two quite separate meanings. One sense is *genetically determined*— 'hard-wired' in a fashionable jargon. The other sense—the only sense in which actual measurement can meaningfully be made—is statistical, not genotypical.

I begin with the genetic version. What would it take to establish rigorously the reality of genetic differences in IQ between races? A lot, it turns out. Only a very elaborate experiment—a real thought experiment—could do it.

We lack now and for the foreseeable future detailed understanding of how the human genotype produces the human brain. Nor do we know exactly how the brain works to produce human behavior. We cannot therefore map out step by step, in detail, how genes-build-the-brain-makes-behavior. Therefore, to understand the genetic determination of IQ we are forced to resort to experiments with whole human infants. To address the black-white issue in a completely rigorous way, we would need to begin with two fetal genotypes, one 'black' and one 'white.' Next, we would need an indefinite number of identical copies—clones—of each genotype. Then, each clone would be exposed to a different rearing environment. (Have we tried out the full range—all relevant environments? How would we know?) Then, perhaps 16 years later, we would give IQ tests to each child. We would get a distribution of IQs for each genotype. We could then ask: how do these distributions differ? Are their means the same or different? Which is higher? Is one distribution more variable than the other?

Suppose we find a mean difference? Does that settle the issue? Well, no, not yet. We are talking about *race* differences here not differences between individual genotypes. A race is a set, a range, of genotypes. So, we need to repeat this impossible process with an indefinitely large sample of 'white' and 'black' genotypes (there will, of course, be debate about which genotypes go into which group). Only after we have this two-dimensional array of genotypes versus IQ can we compare them and come up with a valid conclusion about race difference in IQ. In short, in the absence of full understanding of how genes affect the development of the individual brain, getting a definitive answer to the genetic black-white-IQ-difference question is essentially impossible.

How about the statistical version of the IQ-heritability question: is that answerable? Well, yes, but the term 'heritability' now takes on a very different and much more fluid meaning. The picture shows a relatively simple way to think about the problem. Imagine that each dot is a plot of the average (midpoint) IQ of two parents against the IQ of one of their children. The fact that these points are clustered along a best-fit line, shows a *correlation* between the IQ of the parents and the IQ of their child. For these points, the correlation,[9] h^2, is 0.8, which is in fact within the heritability range found for human adult IQ.

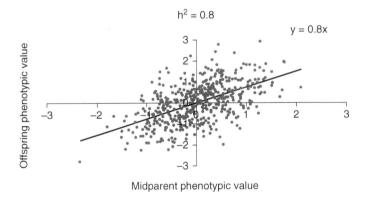

Source: Visscher, P. M., Hill, W. G., and Wray, N. R. (2008) Heritability in the genomics era—concepts and misconceptions. *Nature Reviews*, 9, p. 255. Reprinted with permission.

Is this number a good, stable measure of the real heritability of IQ? Well, no, not really. Statistical heritability has several peculiarities. It is a ratio, of two variabilities: variability of parental IQs divided by the variability of offspring IQs. If either of these numbers changes, so will heritability. It is easy to see how this might happen if we look at extreme examples. Imagine, for example, a very prolific family with many children. What will be heritability for just this population of children? Well, zero, since the denominator, offspring variance, will be greater than zero but the numerator, parental variance, will be zero.[10]

Another comparison is identical (monozygotic) versus nonidentical (dizygotic) twins. For monozygotics the genetic variance between the offspring is zero but there will be some phenotypic variance. To get a meaningful value for h^2, mono- and dizygotic twins need to be compared, "[T]hey ... measured the IQ of 48 pairs of monozygotic, or identical, twins, raised apart ... and 40 pairs of such twins raised together. The dizygotic twins were 69% similar in IQ, compared with 88% for monozygotes, both far greater resemblances than for any other pairs of individuals, even siblings."[11] Much intelligence is genetically determined although it is hard to say exactly how much—or what would even constitute a good quantitative measure. To add confusion, other data show that environment seems to be more important to the IQ of socio-economically deprived children than children reared in middle-class homes. The statistical heritability of IQ also increases with age (the so-called *Wilson Effect*): people resemble their parents more as they get older. This seems paradoxical since the environment obviously has more time to affect our behavior as we get older, but it does give a boost to those who stress the importance of genes to IQ.

The unreliability of statistical heritability as a measure of genetic determination is clearer if we shift away from IQ to more tangible phenotypic characteristics. Imagine a set of siblings who all look completely different. The statistical heritability of these differences will of course be zero, as I have already described. But if the species in question is not humans but cats, and if the feature in question is not IQ but coat color, it is obvious that statistical heritability is an unreliable measure of genetic determination. No one attributes coat color to the environment; it is indeed 'hardwired.' There is no doubt at all that the coat-color differences—which can be quite large in cats—are genetically determined despite their low statistical heritability.

Finally, there is the *Flynn Effect*: Both verbal and nonverbal IQ have *increased* from 1930 to the present. Tests are periodically re-written because performance on the old test gradually improves over the years, so the questions must be redone and re-normed. Are humans really born smarter now than previously—is there natural selection? Or is something simpler happening? Improvement in IQ has been greater in regions like sub-Saharan Africa that began from a lower base. The effect seems to be slowing or even ceasing in some high-IQ countries. James Flynn believes his eponymous effect is because people in the developed world are now exposed much more than previously to tasks that require abstract reasoning and classification, the kind of thing tested by IQ tests. Agrarian societies required less abstraction and more concrete skills. In other words, he argues, it is an effect of environment on each successive generation, rather that some kind of natural selection acting upon the human genotype. Whatever the ultimate cause, the Flynn effect, the Wilson effect, and the effects of environmental variability underline the fact that neither IQ nor its statistical heritability are truly stable measures.

h^2 is nevertheless very useful for selecting breeding cattle and the like. But it seems to me almost useless as a way of resolving issues of human difference, especially for something as much affected by environment as the kind of behavior we call 'intelligent.' In any case, when comparing groups—'races'—what really matters is *adaptability*: how adaptable are children from one group when reared like children from the other? The ability to adapt to a changed environment requires *learning*. Instinct, a fixed reaction to a new situation, is unlikely to help. So, it is important to remember that heritable behavior, including whatever is measured by IQ, is not necessarily instinctive, is not *ipso facto* unmodifiable. Language is the best example of a learned behavior that also has high heritability. Language is 100% learned and 100% heritable—most kids learn the language of their parents. *Accent* on the other hand is much less heritable: kids tend to learn the accent of their peers. Both are learned, but one is (statistically) heritable and the other is not. So much for heritability as a measure of adaptability.

High statistical heritability says almost nothing about how modifiable a trait is. High IQ may help people to adapt to new conditions. To this extent it is helpful. Or it may not: a sedentary and solitary Mensa member is unlikely to do well in a hunting group. What matters for public policy, therefore, is not intelligence so much as people's ability to adapt to changing circumstances—like the onrush of technological modernity which is disrupting so much. To adapt, if not as adults, at least as children. Adult IQ, heritable or not, is not very relevant to this question.

Social and Cognitive Psychology

Chapters 2 and 3 discussed problems with the experimental methods of social psychology (the Tea-Party experiment, for example). This section looks at problems of theory in the 'softer' areas of psychology. How to deal with what is going on inside our heads: this is the enduring issue. With Radical Behaviorism at one extreme, denying the legitimacy of any speculation about what goes on in between stimulus and response, and a discredited Introspectionism worrying about little else, 'scientific' psychology in the early 1960s was in a state of crisis. The following quote from Ulric Neisser's *Cognitive Psychology*, a sourcebook for the 'cognitive revolution,' describes the *zeitgeist*:

> If we do *not* postulate some agent who selects and uses the stored information, we must think of every thought and every response as just the momentary resultant of an interacting system, governed essentially by *laissez-faire* economics.[12]

It is an extraordinary statement. *Quis custodiet ipsos custodes*, 'who shall guard the guardians' is an ancient Roman warning about the dangers of authoritarian political rule. We need a similar maxim for psychology: 'who controls the controller.' Neisser proposes an 'agent' who 'selects and uses' stored information. But what guides the agent? How does he/it select, and why? Surely, we need to know at least as much about him as about his data files? It is another version of the 'intelligent design' problem discussed in Chapter 1.

Later workers attempted to fill in some details. In Alan Baddeley's memory model, the less anthropomorphic term 'central executive' was substituted for 'agent' and some properties were provided for it. But its autonomy was preserved: "It can be thought of as a supervisory system that controls cognitive processes and intervenes when they go astray."[13] On the one hand, dividing the psyche into two parts—one, the central executive, whose workings are obscure, and the other: boxes like 'episodic buffer' and 'phonological loop'—is a reasonable research strategy. It separates the bits about which we can learn something now from the bit we hope to tackle later. On the other hand, by omitting executive function, the model leaves a huge gap where choice and motivation should be. More importantly: it leaves in place the idea that there *is* a somewhat mysterious internal agent who makes all the real decisions.

The obvious alternative is that the brain/mind *is* some kind of *self-organizing system*. One model for such a system is the economists' *laissez-faire* free market, which is dismissed by Neisser, possibly for political reasons (few social-science academics are in favor of market forces). But there are other kinds of self-organization. Bird flocks wheel and dive in perfect synchrony with no market and no sign of an executive. The societies of social insects like ants and bees, function very well without any leader (indeed the great Russian novelist Leo Tolstoy thought the beehive an apt model for an ideal society):

> Give a colony of garden ants a week and a pile of dirt, and they'll transform it into an underground edifice about the height of a skyscraper in an ant-scaled city. Without a blueprint or a leader, thousands of insects moving specks of dirt create a complex, sponge-like structure with parallel levels connected by a network of tunnels. Some ant species even build living structures out of their bodies: army ants and fire ants in Central and South America assemble themselves into bridges that smooth their path on foraging expeditions, and certain types of fire ants cluster into makeshift rafts to escape floods.[14]

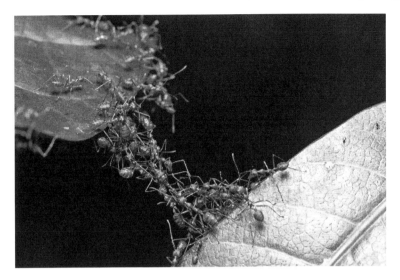

To speed their foraging excursions, army ants build bridges with their own bodies, allowing others to race across a gap.

Source: Shutterstock.

No psychologist disputes the fact that behavior is produced by the brain and that the brain consists of millions of individual cells—neurons, astrocytes, glia, and the rest. No controller is to be seen. Neuroscientists accept the reality of self-organization at that level. We also accept it at the level of the internet, which functions without any central guidance.

The 'Self-System'

Self-organization on the scale of these examples involves thousands or even millions of individual units—birds, ants, neurons, smartphone users. Psychologists would like to find something a bit simpler, a few interacting units which can perhaps be studied and understood separately. The boxes in Alan Baddeley's memory model are one approach. Another is eminent social psychologist Albert Bandura's *self-system*. The idea was first proposed many years ago, but it still finds a place in psychology textbooks. It is designed to avoid the need for some kind of controlling agent:

> A self system within the framework of social learning theory comprises cognitive structures and subfunctions for perceiving, evaluating, and regulating behavior, *not a psychic agent that controls action.* The influential role of the self system in reciprocal determinism is documented through a reciprocal analysis of self-regulatory processes[15]
>
> *[my emphasis].*

If we define a theoretical construct by what it does and how its effects are measured, then it is hard to see much difference between the self-system and

Bandura's disavowed 'psychic agent.' After all, the self-system does 'perceive, evaluate, and regulate' which will look like 'control' to many. ("The self system … allows people to observe and symbolize their own behavior and to evaluate it on the basis of anticipated future consequences" says a recent textbook.[16]) The difference between the self-system and a central executive seems to be that Bandura unpacks his idea bit; it isn't just a black box. We will see just what it is in a moment.

The self-system also goes beyond input-output behaviorism. Bandura argues against the idea of the environment as: "an autonomous force that automatically shapes, orchestrates, and controls behavior." He associates this view with Skinner's *radical behaviorism*, a philosophical position that dispenses (albeit somewhat inconsistently[17]) with any mention of internal processes:

> Exponents of radical behaviorism have always disavowed any construct of self for fear that it would usher in psychic agents and divert attention from physical to experiential reality … it assumes that self-generated influences either do not exist or, if they do, that they have no effect upon behavior. Internal events are treated simply as an intermediate link in a causal chain. Since environmental conditions presumably create the intermediate link, one can explain behavior in terms of external factors without recourse to any internal determinants. Through a conceptual bypass, cognitive determinants are thus excised from the analysis of causal processes.[18] [The surgical metaphors are probably not accidental!]

So, what does the self-system allow us to do—to explain—that 'psychic agents' or the 'intermediate links' implied by behaviorism do not? Bandura's answer is in the diagram. It shows three modes of interaction between the person, P, and

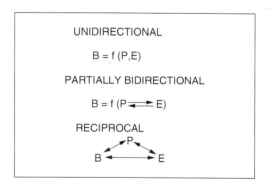

Schematic representation of three alternative conceptions of interaction. B = behavior, P = cognitive and other personal factors, and E = environmental events.

Source: From Bandura, A. The self system in reciprocal determinism. *American Psychologist*, 33, p. 345. Copyright 1978 by the American Psychological Association, Inc. Reprinted by permission.

environmental factors, E. The modes are unidirectional: behavior, B, depends just on P and E; then partially bidirectional: B depends on P and E but each depends on the other; and finally, reciprocal: all three depend on one another. He favors Option 3: reciprocal interaction.

Critics[19] soon pointed out a serious omission in Bandura's scheme: *time* has no place in it. It is entirely static, even though the kinds of interaction it describes necessarily take place over time. Bandura later agreed that things happen over time, but left his original scheme unaltered.

This discussion is very abstract. What are 'cognitive and personal factors'? Just what 'environmental events' are having effects? As always, the problem can be understood better by looking at an example. The example Bandura cites is a naughty child who misbehaves (acts aggressively), then afterwards is punished by a parent. Will the child desist? Bandura suggests that whether he does so or not must depend on his *expectation*: will the punishment continue until he stops, or will it soon cease and let him get away with misbehaving?

There are at least two ways to think about this situation: Bandura's static verbal/diagrammatic scheme, or some kind of dynamic model. Bandura argues that *expectation* is the key: the kid will persist if he expects to prevail but will desist if he does not: "aggressive children will continue, or even escalate, coercive behavior in the face of immediate punishment when they expect persistence eventually to gain them what they seek." What is this *expectation*? Is it conscious and cognitive (as many will assume)? Or at least partly unconscious? Does it matter?

Probably not. What matters is how expectations are formed. Bandura says essentially nothing about this, and in his response to criticism he doesn't even mention expectation. What he does say by way of elaboration is "the interactants in triadic reciprocality work their mutual effects sequentially over variable time courses." In other words, he acknowledges that his three processes interact over time: *history* is important. Perhaps history—the process by which expectation develops—is what matters? No scheme that fails to take history into account can hope to explain even so simple a situation as this.

The next picture shows one kind of historical explanation that makes no use of a self-system. Perhaps it treats internal events as "an intermediate link" as Bandura suggests? Or perhaps it just postulates a hypothetical variable we can call 'expectation' if we wish—but with no assumption about conscious experience. Or, again, perhaps it doesn't matter. *Discipuli picturam spectate* in the words of my first Latin primer: look at the picture. Then I will explain how it was derived.

The picture shows three curves as functions of time. The heavy, solid line shows the level of aggression following two provocations (I'm assuming that the misbehaving child doesn't act aggressively without some provocation). In both cases the provocation causes a spike in aggression which then declines. The curved line underneath represents punishment, which lags after the aggression curve: punishment follows aggression. As the punishment increases, the aggression diminishes; as aggression declines then so does punishment, but with a lag.

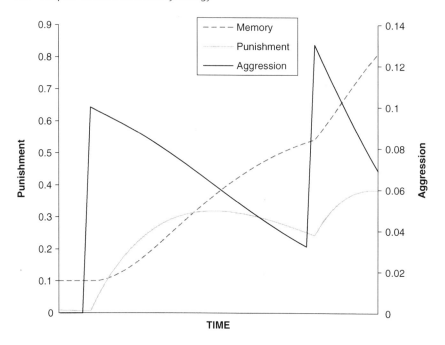

Both punishment and aggression are measurable 'third-party' variables. The dashed line is something different. It is a hypothetical *state variable* (memory) that fits Bandura's idea of 'expectation of punishment.' It rises after each aggressive episode and persists at the higher level.

Is this 'expectation' perhaps responsible for the faster drop in aggression after the second spike? This hypothesis seems to be supported in the next picture which shows four episodes. The peak level of aggression declines during the last episode and dissipates faster and faster across the four. Punishment also declines and expectation (memory) continues to rise.

These events—aggression, punishment, and expectation—occur in the right order in the graphs. But as we learn early on in science, *post hoc* is not necessarily *propter hoc*. The fact that B follows A does not mean that A causes B (correlation is not causation—again) (see the image on p. 67).

So far, these pictures are no more than a graphical depiction of Bandura's example: the naughty kid's behavior and his account for it in terms of expectation of punishment. In fact, the curves were generated by a very simple *dynamic model*. The model does not fit neatly into any of Bandura's three static schemes. It is a *process* which describes exactly how variables representing the levels of aggression, punishment, and expectation change from moment to moment. Aggression also increases in response to external provocation (i.e., some external stimulus). Each variable depends on one or both of the others. Indeed, because there are three equations, a change in any of the three variables will affect both the others. So, in one sense, the

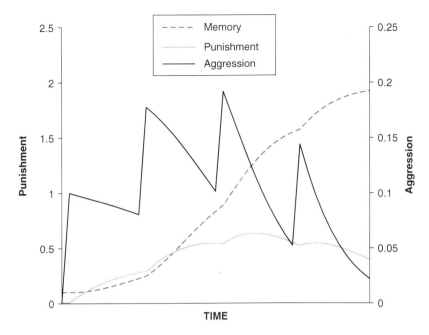

model resembles Bandura's reciprocal interaction option. But there are two important differences: the model is dynamic not static, and it is absolutely specific about how the environmental (x, z) and state (y) variables actually affect one another.

The model that made these curves (three difference equations[20]) embodies the following assumptions:

- Aggression increases in response to provocation.
- When expectation of punishment is zero, the level of aggression declines slowly.
- When expectation is > 0, level of aggression declines more rapidly.
- Expectation increases as punishment and aggression increase.
- When aggression is zero, punishment decreases.
- When aggression is > 0, punishment is boosted.

We can imagine the model as a box (transparent, not black!) with four control knobs setting the four parameters. By twiddling the knobs, we may be able to understand individual differences in aggressive behavior. The model is incomplete and certainly wrong in many ways. But it does three things: it helps to define the theoretical problem; it incorporates time; and it suggests simple predictions—about the effects of increasing or decreasing the level of punishment and of interposing longer or shorter intervals of time between provocations.

Static proposals like the self-system—proposed in 1978 but still alive in textbooks—impede progress. They are phrased in familiar terms that appeal to our intuitions. They seem reasonable. They are difficult to test and to disprove: the self-system, like the ego and id before it, is still around. We are all aware of our *selves*; how nice to see the idea built in to social psychology theory. But, as the critics pointed out, behavior is not static, to be mapped like Greenland, but dynamic, like the ocean with tides and waves. The simplicity of static schemes like the Freudian trio, or Bandura's self-system, has inhibited research on behavior as a moving target. More theoretical progress may be made via dynamic models. That path is tougher than speculation about self-system and the like, but is philosophically sounder and may be more fruitful of testable, useful hypothesis. A path less traveled may lead on to fields more fertile than the current state of social psychology theory, where decades-old bad ideas live on, sustained by their easy familiarity.

Notes

1 See, for example, review by Clifford Geertz of a book by Bent Flybjerg: Empowering Aristotle, *Science* 06 Jul 2001, 293(5527), p. 53: "Flyvbjerg clearly demonstrates that there are models more appropriate to the social sciences than those derived from molecular biology, high-energy physics, the mathematical theory of games, and other up-market, hard-fact enterprises." Also, How Reliable Are the Social Sciences? By Gary Gutting *New York Times*, May 17, 2012 9:30 pm. https://opinionator.blogs.nytimes.com/2012/05/17/how-reliable-are-the-social-sciences/?_r=0

2 Anthony Oberschall: Cooperation or conflict: First encounters of Pacific islanders and Europeans. TORCH talk, February, 2017.

3 E.g., The Dunedin Multidisciplinary Health and Development Study: overview of the first 40 years, with an eye to the future. https://moffittcaspi.com/sites/moffittcaspi.com/files/field/publication_uploads/PoultonMoffittSilva_2015.pdf

4 Plausibly, but wrongly, attributed to Charles Darwin.

5 *Operationism* is a doctrine proposed by US physicist Percy Bridgman in 1927. PB proposed that the *meaning* of a (scientific) concept is equivalent to the actual methods used to measure it. Delphic Austrian philosopher Ludwig Wittgenstein said much the same thing at much the same time, about language: "the meaning of a word is its use in the language."

6 The embarrassing exceptions, of course, are athletes, 'legacies' and students given preference because of their race, who may have lower-than–college-average IQs and do indeed tend to perform worse than other students in difficult courses.

7 Galton, F. (1869) *Hereditary genius.* Macmillan & Co.

8 www.technologyreview.com/s/538431/deep-learning-machine-beats-humans-in-iq-test/

9 *Correlation* can range from zero (a random cluster of dots) to one (all the points lie on the same straight line).

10 Statistically, but not really. Gene recombination allows for differences between siblings (see the discussion of cats, below). But recombination is not part of statistical heritability.

11 Ridley, M. (2012) Is IQ in the genes? Twins give us two answers. *Wall Street Journal.* Available at: www.wsj.com/articles/SB10001424052702304898704577478482432277706

12 Neisser, U. (1967) *Cognitive psychology.* New York: Appleton-Century-Crofts, p. 293.

13 https://en.wikipedia.org/wiki/Baddeley's_model_of_working_memory

14 www.quantamagazine.org/20140409-the-remarkable-self-organization-of-ants/

15 Bandura, A. (1978) The self system in reciprocal determinism. 34(4), *American Psychologist*, 323–345.

16 http://highered.mheducation.com/sites/0072316799/student_view0/part3/chapter11/chapter_outline.html

17 Skinner was happy to talk about 'private stimuli'—which demand of course an equally private entity to receive and react to them: a 'self-system,' 'central executive' or 'agent' in disguise, perhaps? In a 1948 talk about language he also spoke about *latent responses*.

18 Bandura (1978), p. 348.

19 Phillips, D. C., and Orton, R. (1983) The new causal principle of cognitive learning theory: Perspectives on Bandura's "reciprocal determinism." *Psychological Review*, 90(2) pp. 158–165. Bandura, A. (1983) Temporal dynamics and decomposition of reciprocal determinism: A reply to Phillips and Orton. *Psychological Review*, 90(2) pp. 166–170. Staddon, J. E. R. (1984) Social learning theory and the dynamics of interaction. *Psychological Review*, 91, pp. 502–507.

20 In difference-equation form, the model is: $x(t+1) = Ax(t)-Bx(t)y(t)$; $y(t+1) = y(t)+z(t)x(t)$; $z(t+1) = Dz(t)+Cx(t)$, where A, B, C and D are fitted parameters. x = aggression; y = expectation; z = punishment; A =.99, B = .1, C = .5, D = .9.

5

SOCIAL SCIENCE

Economics

It is better to be roughly right than precisely wrong.

John Maynard Keynes

I have observed more than three decades of intellectual regress.

Paul Romer[1]

Economics is the most prestigious and best-paid, social science. Why the prestige? Not because it can predict better than, say social psychology or personality theory. Many mainstream economists failed to predict the 2008 crash in the United States, wrongly predicted inflation supposed to follow the subsequent financial stimulus, and predicted a disaster that did not happen in Britain after the Brexit vote in 2016. Little wonder that some economists deplore any attempt to predict: "[E]conomists probably regret the way that their noble trade has been associated too much in the public mind with forecasting."[2] But of course it is the duty of science either to predict or—as in the case of some parts of quantum physics and biological evolution—show why prediction is not possible. Limited as it is, economics may guide. It should not decide.

Why is economics so influential, despite its very many limitations? It does purport to explain things like job growth and the creation of wealth, weighty matters that affect everybody. Shamans, soothsayers, and the like used to pretend to heal the sick and predict the weather. They were important figures until better methods, in the form of meteorology and medical science, were devised. Clever men, unchallenged and in the absence of effective alternatives, who propose to help with important issues, will always gain a following. Our understanding of how real economies work is as modest as our need is great. Hence almost any confident voice will get a hearing, especially if it appears to be backed by science. But how scientific is economics, really?

The scientific method of economics is largely theoretical: "Models make economics a science[3]" in the words of a respected contemporary economist. My focus is on economic theory. What kinds of theories are there? What is their conceptual basis? Is it sound? Can economic theories predict accurately? Can they be tested? This chapter and the next two deal with existing theories, their problems, and possible alternatives.

Two sourcebooks are Adam Smith's two-volume *Theory of the Moral Sentiments* (1759), which is basically a psychology monograph, and his more famous *Wealth of Nations* (1776), generally regarded as the first book on economics. In TMS, Smith follows what we would now call a folk-psychology view of the springs of human action. He accepts the scientific value of things like sympathy, (what might now be called 'empathy') sorrow, perfidy, ingratitude, dignity and honor, the sense of duty, social affections, and their causes. He comments on effects. Age affects sympathy, for example: "In ordinary cases, an old man dies without being much regretted by anybody. Scarce a child can die without rending asunder the heart of somebody," and yet "The young, according to the common saying, are most agreeable when in their behaviour there is something of the manners of the old, and the old, when they retain something of the gaiety of the young. Either of them, however, may easily have too much of the manners of the other." Smith makes no attempt to explain these attitudes. He simply accepts them as part of human nature.

Are these commonsense terms—'the levity, the carelessness, and the vanity' of youth, etc.—and informal analyses of social interactions useful as science? This is a historical issue in psychology. The Wurzburg school in Germany and Edward Titchener (1867–1927) in the United States tried to use introspective reports of feelings, beliefs, and perceptions as a basis for what Titchener called 'structural' psychology. The effort failed: "In the 'new' psychology's first fifty years, the description of consciousness had resulted in no large interesting systematic body of knowledge."[4] The reasons were many, but perhaps the most important is that not everything—indeed, rather little—that is going on in your head is accessible to consciousness. History suggests that Smith's attempt to use consciously perceived 'moral sentiments' as a basis for psychology is also likely to be a dead end.

Two Kinds of Explanation

On the other hand, Smith has a very modern approach to psychology as a science:

> The sentiment or affection of the heart from which any action proceeds, and upon which its whole virtue or vice must ultimately depend, may be considered under two different aspects, or in two different relations; first, in relation to the cause which excites it, or the motive which gives occasion to it; and secondly, in relation to the end which it proposes, or the effect which it tends to produce.[5]

Ludwig von Mises, a founder of the Austrian School[6] of economics, said much the same thing: "There are for man only two principles available for a mental grasp of reality, namely, those of teleology and causality" that is, in modern terms, human behavior is to be understood in terms of either cause or outcome. This is an approach congenial to many modern psychologists, especially the more behavioristically inclined. Smith's *cause* and *motive* correspond to stimulus and motivational state in the modern lexicon. These are the source of a person's *repertoire*, the set of activities, both actual and potential, that he brings to any situation. The *effect*, Smith's *end,* is of course just the consequence—reward or punishment, praise or blame, indifference or excitement—that follows on any action.

This distinction between two kinds of explanation, causal and outcome-based, is essential to a proper understanding of economics, where outcome-based theory dominates. Causal theories are relatively straightforward: a cause, manipulated in an experiment, produces a reproducible effect. Chemicals in their food produce changes in the songs of starlings, as described in Chapter 3. Outcome-based—teleological or, rather confusingly termed *functional*—theories are quite different. For example, in Chapter 6 I will discuss a simple reinforcement schedule equivalent to the casino-style two-armed bandit: two choices each paying off with a different probability. Animals and people maximize on such schedules. After exploring each option for a while they finally settle on the higher probability. Their behavior can be explained by an outcome-based, teleological—functional—theory of reward-rate maximization. But how do they do it? Economists derive the optimum by equating marginal payoffs, as I describe in a moment in connection with the Marginal Value Theorem. But many experiments show that this is rarely the way that people, never mind animals, solve a problem like this. Do they maximize on the two-armed bandit problem by keeping a mental record of past payoffs? Over what period? How often do they compare records for the two choices? The details make a difference to how the subject will respond if the probability changes and to how he will adapt to changes in the actual time pattern of rewards. A person may behave optimally in one situation but sub-optimally in another, apparently similar, situation. Without knowing the actual underlying process of choice, the fact that he behaves optimally in one situation offers little guide to how he will do in others. Accurate prediction demands an understanding of *process.*

In his pioneering tract of economics, *The Wealth of Nations,* though, Smith makes little use of the psychology he elaborated in TMS. In WN his concern is with what would now be called *macroeconomics*, the study of groups—industries, trades, labor, etc. He sets up the problem in this very modern way: The goods and services a nation gets are either produced by it or imported. The affluence of the nation depends on the ratio of this total amount to the number of people who will buy or use it. Given a certain proportion of workers, their output depends on their efficiency which, in turn depends on the capital—tools, land, skills—with which they can work.

A modern economist might say the same thing more formally. If we define the terms as follows:

Affluence = A
Total goods = G
Goods imported = G_I
Goods made = G_M
Total population = N
Number in work = W
Capital = C
Training = T

Then a highly oversimplified summary would be something like:

$$A = G/N$$
$$G = G_I + G_M$$
$$G_M = f(W,C,T)$$

where f is some mathematical function. Obviously, the task for economists is to understand function f: how do the number of workers, the capital stock available to them and their expertise, all combine to generate goods? Unfortunately, human ingenuity enters in to factors C (capital) and T (training) in ways not easily captured by a mathematical formalism. *Affluence*—wealth—is not the same as wellbeing or the common good; and *number in work* does not distinguish between full or part-time or the type of work. So, it is perhaps just as well for Smith that he did not go down the mathematical road. But many modern macroeconomists do follow precisely this kind of reasoning, though in much a more complex way than this cartoon version. What Smith did do was use as an example an innovation much in evidence in 1776. *Division of labor* was a novelty that was much increasing the efficiency of manufacturing, and Smith explores its implications in some detail, beginning with his famous example, the making of pins:

> To take an example, therefore, from a very trifling manufacture, but one in which the division of labour has been very often taken notice of, the trade of a pin-maker: a workman not educated to this business ... nor acquainted with the use of the machinery employed in it ... could scarce, perhaps, with his utmost industry, make one pin in a day, and certainly could not make twenty. But in the way in which this business is now carried on, not only the whole work is a peculiar trade, but it is divided into a number of branches, of which the greater part are likewise peculiar trades. One man draws out the wire; another straights it; a third cuts it; a fourth points it.

By the end of the 18th century, technological advance had become a major, possibly the largest, force for economic change. How on earth can economic science make sense of this rapidly growing complexity? There are several strategies, each flawed in its own way. (1) Commonsense verbal description, a continuation of Smith's approach. (2) More or less sophisticated mathematization, following Smith's macro lead to its logical conclusion. (3) The individualistic, nonmathematical, axiomatic approach of the Austrian School. (4) And finally, *behavioral economics*, a relatively recent development that promises to link psychology to economics. The idea that economic choices can be explained by individuals or groups choosing so as to maximize their own gain—*rational choice*—has been the foundation of economic thought until relatively recently. Behavioral economics uses experiments with human and animal subjects in contrived situations as a way to unpack the concept of *rationality*. I will say a bit more about rationality and behavioral economics in a moment. First a comment on the ubiquity of mathematics in economics.

Mathematization

> *We find ourselves confronted with this paradox: in order for the comparative-statics analysis to yield fruitful results, we must first develop a theory of dynamics.*
>
> Paul Samuelson[7]

Adam Smith eschewed mathematics as did most economists who followed him, from Karl Marx to Ludwig von Mises. The Austrian School shared a belief in what it called the *aprioristic* character of economics with the mathematical approaches that came to dominance a little later—but without the mathematics. The behavior of interest to economists involves *choice* and *purpose*. From these principles alone von Mises purported to derive an a priori science of choice. I say 'purported' because, without the usual logical apparatus of axioms, proofs, etc., the a priori aspect is hard to discern. The Austrians were inoculated against the mathematical approach by their focus on the individual and suspicion that equations promise more than they can deliver: "… if a mathematical equation is not precise, it is worse than worthless; it is a fraud. It gives our results a merely spurious precision. It gives an illusion of knowledge in place of the candid confession of ignorance, vagueness, or uncertainty which is the beginning of wisdom."[8] The rest of economics happily dealt with groups and embraced mathematization.

By the middle of the 20th century, formal approaches had begun to gain the upper hand. The Chicago School of Economics, led by stars like Milton Friedman, Gary Becker, and Eugene Fama, was a strong advocate not just of free markets and freedom generally, but of predictive mathematical models. Jacob Viner, an early member of the group, defined the new field in this way:

> Neoclassical economics is deductive or a priori in its method. It is static. It does not discuss laws of change, but it discusses processes of equilibrium under given circumstances. It takes value as its central problem and approaches economic problems in value terms, meaning exchange value …

It puts little emphasis on consumption … It does not explore the origin of value … It is an objective description of the process of price determination under certain assumed conditions … The main emphasis is on a descriptive, objective explanation of the way in which prices come to be what they are.[9]

This definition is from 1933, but little has changed since then. It is an extraordinary statement—although, to be fair, Professor Viner apparently finessed the question later, saying: "Economics is what economists do."

Viner's attempt to define the science of economics is both prescriptive and self-contradictory. He describes economics in a way that suggests it is almost independent of the real world, defined without the need of observation, *a priori* indeed! Like most economic theorists after him, he ignored the distinction between functional/teleological theories and causal theories.

General Equilibrium

[T]he powerful attraction of the habits of thought engendered by 'equilibrium economics' has become a major obstacle to the development of economics as a science.
Nicholas Kaldor[10]

Here are some problems with Viner's claims: "It [economics] does not discuss laws of change, but it discusses processes of equilibrium." But how does economics *know* that equilibria exist, that "laws of change" are not necessary? George Soros, a brilliant investor, not a fan of the efficient-market theory, and far from a hero to the economics community, simply dismisses the idea that markets are self-stabilizing: "Because markets do not tend toward equilibrium they are prone to produce periodic crises."[11] How do we know that market prices, for example, are not in constant flux—like, say, the weather (albeit less predictable) or waves on the ocean, but on a longer time scale? And why neglect consumption? How does Professor Viner *know* it is unimportant to price? Indeed, how does he know that his version of economics is relevant to the real world at all?

The paragraph is also self-contradictory: it "takes value as its central problem" but "does not explore the origin of value." Why not? The huge changes in the value/price of things which we call *fashion, contagion,* and the *madness of crowds*, show that value is not a constant. Economics is 'deductive or a priori' but also 'descriptive and objective.' Well, which is it: *descriptive*, in which case it depends on observation, or *a priori*, in which case it does not!

In the 1950s, several brilliant economists[12] did in fact address the question of market equilibrium. But they didn't do as Samuelson suggested and come up with a dynamic theory. Instead, they proved that an equilibrium *must* exist—if the system obeys some simple rules. The rules involved preferences—they must be *convex*, which roughly corresponds to diminishing marginal utility, which I will get to in a moment. *Competition* must be 'perfect'—actually not so simple to define or test. And finally, there is *demand independence*—the demand for tea must be unaffected by, for example, the demand for coffee. Economists in general were undeterred by the unrealism of these assumptions. Despite intermittent criticism, most economists took away the idea that equilibria are more or less guaranteed. Professor Viner's confident and implausible claim that economics as a discipline is almost as certain as Euclidean geometry was in fact in the mainstream. (We'll see in Chapter 8 that there is a sense in which Viner was correct, but only at the expense of economics' scientific credentials.)

Economists are invariably faced with bewildering complexity. They need to simplify. They have done so by coming up with reasonable-sounding, but often unverifiable, formal models. How well does the mathematical/graphical approach do as a way to understand, for example, price fluctuation? It does well in one way, not so well in another. Rational economics, in the form of optimality theory, is always useful in directing attention to *relevant variables,* features of the environment that are likely to affect whatever it is you are interested in. For example, suppose there is a very bad 'weather event' in Sri Lanka that severely damages the tea crop: what will be the effect on tea prices in the United States? Rational economics suggests several useful questions: where else is US tea imported from? What are their crops likely to be? Are there substitutes for tea—will many people be happy enough to drink coffee? What is the state of the coffee market—did production go up or down this year? What is the state of the general economy—are incomes increasing or declining? All these are things that may be helpful in guiding action, in buying or selling tea and its substitutes.

So far, so good; answers to questions like this are surely vital to producers, consumers, and policy makers. But science should provide answers as well as questions. A mature economics should provide not just a list of potential causes, but a measure of their individual and collective effects. Yes, availability of a substitute for coffee will affect demand for coffee when its price changes, but by how much and through what process? How fixed is people's love of tea and coffee? How are price and purchase decisions actually made? Some of these questions are about psychology, others about modes of production, still others about politics. Ultimately almost any macroeconomic question involves human society in general. It is probably unfair to expect any theory, least of all, perhaps, an elegant mathematical theory, to explain such a multiplicity of causes and effects. But correctness should not be sacrificed to elegance. In economics, all too often it is.

Supply and Demand

Economics is perhaps best thought of as a set of questions generated through some form of optimality analysis: as pertinent questions, rather than quantitively defined answers. If we know what things *should* be important to consumers, perhaps we will be better able to understand what consumers actually do. But a set of questions is not a science, no matter how compelling the formal scheme from which they are derived. Indeed, a scheme's very elegance may conceal the limited understanding on which it is based. The analysis of supply and demand, one of the most successful economics paradigms, is a good example.

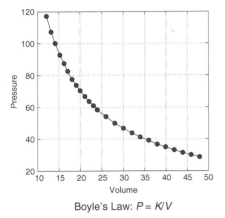

Boyle's Law: $P = K/V$

Here are some problems raised by mathematization. Economics texts are full of graphs, but I begin with a graph from physics: Boyle's Law says that at constant temperature, pressure and volume of a fixed amount of gas are inversely related. In an experiment, P and V can be directly measured; V can be altered and the new P measured; P and V do not vary spontaneously. The law is easily verified.

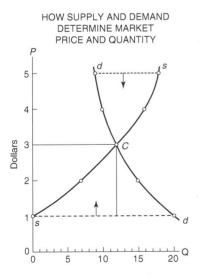

HOW SUPPLY AND DEMAND DETERMINE MARKET PRICE AND QUANTITY

The 'equilibrium' price is at the intersection of the two curves, where the price is such as to make the amounts demanded and supplied exactly equal.

Source: From Paul Samuelson, *Economics: An Introductory Analysis* (McGraw-Hill, 1955).

The next graph is almost as simple. This example is from a famous economics textbook. It shows how the number of bushels of corn produced each month is related to the price per bushel. It illustrates one of the most fundamental economic principles: the *law of supply and demand*. The graph works like this:

> At a higher price [*y*-axis], the dotted line shows the *excess* of supply [*s*] over demand [*d*]. The arrow points downward to show the direction in which price will move because of the competition of *sellers*. At a price lower than the equilibrium price, $3, the dotted line shows that demand overmatches supply. Consequently, the eager bidding of *buyers* requires us to point the arrow indicator upward to show the pressure that they are exerting on price. Only at point C will there be a balancing of forces and a stationary maintainable price.[13]

This seemingly obvious little diagram raises questions—questions that can be answered satisfactorily for Boyle's Law but not for the law of supply and demand. How do we know the form of the two curves? If you check a selection of textbooks you will see that all that the graphs share is that one curve goes down and the other up with price. The scale and actual form are arbitrary. How fixed are these two curves: that is, do they change from day to day, or even moment to moment, or are they stable as most theorists assume? Most importantly: do they represent accurately the process that underlies price? By *process*—von Mises' and Adam Smith's *cause*—I mean the actual real-time, moment-by-moment mechanism that drives price changes.

First, unlike the pressure-volume curve, neither the supply nor the demand curve in the picture can be directly measured. Economists seem to assume that supply and demand curves are both real and *causal,* even though the supply-demand account is what von Mises would call teleological and many other social scientists call *functional*. Economists sometimes try to discover—identify—supply-demand curves from empirical data. Unsurprisingly, the results are usually inconclusive: "The treatment of identification now is no more credible than in the early 1970s but escapes challenge because it is so much more opaque," says well-known economist Paul Romer in his *The trouble with macroeconomics.*[14]

Second, we do not in fact know how stable supply-demand curves are, or even if they have an existence independent of the graph. They conform to intuition; that's the best we can say. The graph looks like mature science, but is not. It is a visual metaphor, like much of 'comparative statics,' as this branch of economics is termed. The graph says no more than this: when prices rise, demand tends to fall and production to rise.

But putting all into a graph promises much more. It suggests a mechanism by which the equilibrium is reached. What might that mechanism be? Professor Viner, perhaps wisely, "does not discuss laws of change." But there *is* a process, of course. Price is set by hundreds or even thousands of producers and consumers, each making a decision according to his own individual psychology, a psychology that is still little understood at this level. There is a process, but it is very difficult indeed to know how to capture it in a formalism, which is why most economists avoid the problem altogether.

Samuelson had a stab at dynamics in an early mathematical paper.[15] For example, about the dynamics of supply and demand he proposes as a starting point the difference between supply and demand curves:

$$\frac{dp}{dt} = H\left(q_D - p_S\right),$$

but without defining q and p other than as variables related (presumably) to supply and demand. The equation just says that the rate of change of price, dp/dt, is related to some function, H, of the difference between some variables, q_D and q_S, related to supply and demand. But without knowing all three of these entities precisely, the equation says nothing more than price depends on the difference between supply and demand. But it does so at the cost of baffling, or perhaps overly impressing, the novice reader.

Nevertheless, Samuelson's later textbook description of the supply-demand graph hints at a relatively simple process, a process that does not require individual psychology: 'pressure' is exerted on price, which moves up or down accordingly; there is a 'balancing of forces.' Let's take him literally and see how this balancing might work.

The next graph shows an S-shaped curve called the *logistic function*. Number—of bacteria, for example—on the Y-axis—grows at first slowly with time, but at an accelerating pace; then it slows down, approaching a steady maximum. This curve idealizes the way that populations of biological organisms grow over time. At first the growth is Malthusian, geometric: time 0: number N; time 1: number rN, time 2: number r^2N, where $r > 1$ is the growth rate. But eventually the population reaches the limit of its habitat and stabilizes at a maximum, usually symbolized by a second parameter K. The full equation is $\delta x = rx(K-x)$, where δy is the change in population x from instant to instant, r is the exponential growth factor, and $(K - x)$ limits the growth of x to a maximum of K. (Hence the terminology in biology: K-selection means few offspring, much nurturing, like people and elephants; r-selection, many offspring, little nurturing, like turtles and aphids.) When $x << K$, the exponential term rx dominates; when x approaches K, growth is limited by the term $K - x$.

The logistic curve looks like a fair approximation to the way that economists imagine that production grows when a product is priced well above production cost. At first rapidly, but then slowly as production cost and price come into equilibrium.

The next picture shows that this expectation is more or less correct. It plots two similar simultaneous logistic difference equations.[16] The change in price is inversely related to the change in supply: if supply is increasing, price will decline and vice versa. Similarly, the change in supply is proportional to the difference

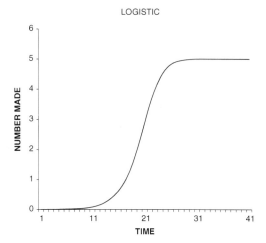

between current price and the cost of production: the larger excess of price over production cost, the faster production grows.

The two curves look like the supply and demand curves in the previous picture, but they are of course quite different. Both are functions of time and could in principle be directly measured. One shows how the number produced (in arbitrary units, per unit time) changes over time; the other shows how price changes. The *initial conditions* are a high price (set to 1 for convenience) and a low supply (NUMBER close to zero). From that point, in response to the high initial price, the number produced increases. At the same time, price decreases in response to the increasing supply. The process stabilizes, the lines flatten out, when price is equal to production cost (0.1, in this case).

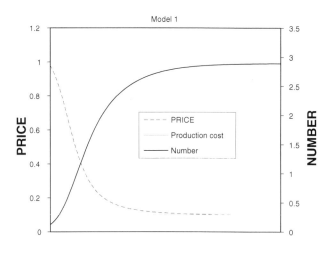

The next picture shows how conventional supply and demand curves are used to explain the effect of changes in supply or demand. The picture shows the effect of an increase in supply. Given these monotonic supply-demand curves, price will decrease. A decrease in supply will obviously have the opposite effect. But the diagram adds little to the simple statement that supply and price vary inversely.

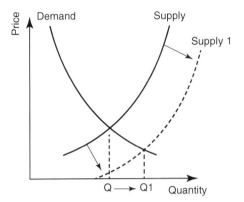

It is interesting to see what a comparable change—in cost of production, hence price—does in the dynamic model. The next picture shows the result for three cases. The sloping straight line at the bottom of each graph is the cost of production: slowly decreasing or increasing, or increasing more quickly. The first graph looks sensible: as cost of production declines, supply goes up and price decreases until price and cost of production come together. By the generous standards of economics, this looks like an OK model: it shows that the process does equilibrate and that the equilibrium is, as expected, at a price equal to production cost.

But the next two graphs blur this pretty picture. The model and its parameter values are unchanged. But now in both cases the increase in cost produces an up-and-down change in both price and supply. The large increase in production cost in the last picture apparently shows price falling below production cost by the end of the graph. In other words, a slow, steady change in one parameter, production cost, causes changes on both supply and price that are not gradual at all. The

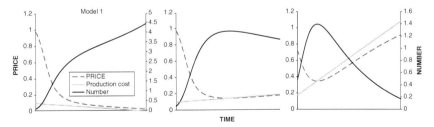

dynamic model shows (why did it even need to!) that sharp changes in things like price and supply can occur without any identifiable, external 'shock.' Dynamic processes, even very simple ones, often behave in surprising ways.

The unexpected results shown in the picture raise an obvious question: when production cost rises, are changes like this observed in actual markets? I don't know; I leave it to empirical economic research to collect data like that. But the dynamic analysis does raise serious questions about a static approach, which is plausible more because it matches our own unconscious assumptions about the essential stability of market processes than because of the accuracy of its real-world predictions.

So, just how much can we expect from existing research methods, theory especially, in economics? And if current approaches are flawed, what are the alternatives?

The answer to the first question is not encouraging. If even such a simple dynamic model as this can behave in unpredictable ways, how likely is it that reality will be more predictable, rather than less? What reason do we have to believe that real supply and demand will behave in the precise way suggested by the standard, static supply-demand graph?

"Very little," I suggest. In support of that skepticism I show one more graph, from a slightly more complicated version of Model 1. Model 2 has three parameters rather than two. With the right choice of parameters, it has almost the same response as Model 1 in the first two pictures: it equilibrates when cost is constant, and shows an up-and-down response to changing cost, like the second two pictures above. But for certain parameter settings it shows the kind of chaotic behavior in the next picture. Price and number vary in an unpredictable manner from moment-to moment, even though nothing in the environment is changing.

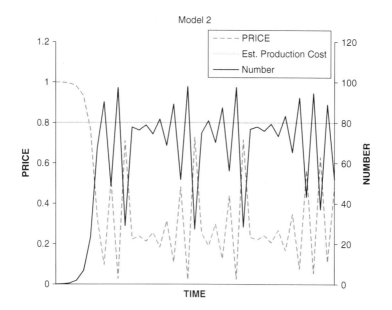

The take-away is that price and output can vary without any external cause at all, even if the system is entirely deterministic. A static, model of supply and demand simply misses the point.

If I were to take the dynamic model seriously, then it would be necessary to identify the parameters. They would have to represent things like the level of technology, union activity among workers and the political climate. These things can obviously change over time. Hence, if a dynamic model something like one of these actually describes the real-world supply-demand process, we should not be surprised to see prices switch unpredictably from stability to chaos in the absence of obvious environmental triggers

There is another way to deal with variability in data given a static model that shows none: add a 'noise' component. This is a common, but dangerous, practice. Unless the modeler can be sure he has captured every deterministic element, adding noise just papers over the problem. Best to begin as I have here, with a completely deterministic model and see how far you can get before adding cosmetic noise. Apparently, you can get pretty far, since even quite simple processes can behave in complicated ways.

So, lesson number one is this: sudden change in an economic variable like price need not reflect an equally sudden cause. Economic numbers can vary all on their own with no forcing stimulus or 'shock'—just like the weather. Economists will never understand processes like this unless they ignore Viner's dismissal of 'laws of change.' If in fact change is what it is all about, we would do well to study not equilibrium, but variability.[17]

Sometimes variability *is* caused by external factors, of course. One such factor is time delay. Some products, like computer games or recorded music, can respond immediately to increases in demand. But others, like real estate or particular crops, absolutely cannot. In real estate, for example, if prices seem to be increasing, developers will begin new projects. Will they begin too many, so that supply will soon exceed demand? Or too few, so that price will continue to rise. A developer must begin a project based not on current demand but on his estimate of demand at the time he is ready to sell. No one really knows either exactly when that will be or what the demand will then be. It is therefore hardly surprising that markets for real estate and agricultural products often show large unaccountable price swings.

To try and apply static, time-free models to such systems makes no sense at all. This was in fact recognized quite early on in the form of the *cobweb model*.[18] The basis was still the supply and demand curves, which were assumed to be fixed. But introducing a time lag between price changes and production decisions yields an oscillating pattern of price changes. This simple stab at dynamics could have led to more interesting dynamic theories but, as far as I can tell, did not.

There are also very practical implications of the insight that real economics is not static: simple, unqualified, predictions are rarely warranted. An example: a change in law that introduces, or increases, the minimum wage is perhaps the closest thing in economics to an actual experiment. The "AB design" (no minimum → minimum) is imperfect, for the reasons discussed in Chapter 1 (ideally, B should be repeated). And the timing of the change is unlikely to be random, as good experimental design would require. Minimum wage is more likely to be increased when the economy is booming than when it is depressed. If most wages are already higher than the minimum wage, it will obviously have little effect on either wages or employment. The contingencies[19] are complex, but the prediction of standard economics is not: increasing the cost of labor by raising the minimum wage must decrease demand for it. Minimum-wage increase is therefore bad because it will always cause unemployment. Yet the empirical evidence is mixed.[20] There is as yet no agreement on whether this most elementary economic prediction is in fact false, correct sometimes or, possibly, always true under specified conditions.

Polymath English intellectual John Maynard Keynes was perhaps the most influential economist of the 20th century. He was sophisticated mathematically—he wrote a long book on probability—but reasoned commonsensically. He was instrumental in designing the Bretton Woods currency arrangements that stabilized international finance after the end of the Second World War. He made much money for the endowment of Kings College, Cambridge, of which he was for some years Bursar. He argued that government debt could save an economy in recession. He was part of the sexually flexible *Bloomsbury Group* of socialist intellectuals in London in the inter-war years, a group that included writers Virginia Woolf and E. M. Forster, handsome artist Duncan Grant and, on the periphery, philosopher Bertrand Russell. Keynes was happily but improbably married to strong-minded Russian ballerina Lydia Lopokova from 1925 until his death just after the war.

Keynes thought that 'noise'[21] might be intrinsic to all economics. He famously spoke of 'animal spirits,' by which he meant the built-in caprice of humans who will sometimes change their habits and preferences for no apparent reason. Add to animal spirits the distinctively human tendency to follow others, either fashion icons or market gurus. Young women these days seem to favor deliberately torn blue jeans. These garments, which look as if they have been salvaged from the trash, are now produced in quantity and command a price. What happened? Pop music icon Beyoncé Knowles sometimes wears torn jeans. Is she responsible for the fad? Mechanical wristwatches, briefly eclipsed by more accurate quartz ones, now command a much higher price.[22] Why? Who knows. Whatever the reason, this kind of unpredictable, not to say incomprehensible, contagion is notorious in the fashion and art markets. It conforms to no simple supply-demand analysis, but it affects markets in major ways.

Simple dynamic models do not deal with these complicating factors. But the models show that even without contagion and caprice, we can still expect price and production to show variation. The dynamic models I've discussed are obviously far from adequate, but that's not the point. The point is that *static models don't work at all*. Economists cannot ignore time and hope to gain any insight even into something as basic as supply and demand. The visual metaphor of supply-demand curves far from being a help, in fact impedes understanding by assuming away the real problem: change over time. How should the models be modified to take account of time delays? Of contagion and animal spirits? Is there a better dynamic approach? The standard static metaphor is no help. But if economists began to think in dynamic terms, attending to rather than neglecting 'laws of change' they might at least see where the gaps are.

Rationality and Marginalism

All agree that the idea of *value—utility—*is at the heart of economics. Some, like Professor Viner, decline to inquire as to its source. Others are happy take value as *sui generis*, a fact to be accepted without explanation—like the fact that intentionally torn blue jeans have a nonzero price. Still others speak of *revealed preference*, the idea that the value people attach to things can be measured by their choice behavior. Even the Austrians, outliers to mainstream economics, agree that teleology, the study of final causes, of *purpose*, defined by the end desired, is key to economic behavior.

That economic behavior can be explained by *utility maximization* is one of the least-controversial ideas in economics. Maximizing total utility is the economist's definition of rationality:[23] people act rationally to maximize their welfare. Behavioral economics (about which, more in a moment) has put a dent in this idea, but it is still the dominant theme in economics. Here are the basic assumptions:

- People and animals attach a certain value to every available choice option. Economists call the value so assigned, the *utility* of the option.
- Utility can be entirely hypothetical, its properties chosen so as fit a particular model; or it can be measured through experiment, in which case it is called *revealed utility*.
- Utility usually shows *diminishing marginal returns*: if the first slice of pizza is worth (to the consumer) X, the second slice will be worth X-Δ, the third perhaps X-Δ-δ, where Δ and δ are small positive quantities. In other words, each incremental bit of pizza has less utility than the one before.
- People (and perhaps animals) act so as to maximize utility.

Marginal Value Theorem

Here is a simple example from biology. Birds forage for food that is not evenly distributed. Usually the bugs they seek occur in *patches*, trees or bushes with prey in high concentration, separated by ground with no prey. What is more, as the bird forages on a bush, he will find it increasingly hard to find each new bug as he depletes the supply. Even if every bug is worth the same, has the same utility, foraging in patches shows diminishing marginal returns over time.

Many years ago, behavioral ecologists asked an obvious question: how long should the bird work a depleting patch before giving up and looking for another? Suppose, for simplicity, that travel time is constant and each patch is identical and depletes in the same way. The picture shows this idealized version. The curve on the right shows how the bird's intake grows more and more slowly with time. The distance labeled "Transit time" shows the time it takes to travel between patches. The slanted dotted line shows when the bird should quit a patch and move on: when its *marginal rate of eating*—the expected rate of eating over the next brief interval of time—is equal to the overall rate of eating, including transit time (indicated by the slope of the dotted line). In other words, the bird can maximize its food intake by equalizing the marginal rate of eating for both choices: staying or leaving the patch. Maximizing utility is achieved by the equalization of marginal rates. The bird's behavior is *rational* according to economists' usual definition.

Real food patches vary in size and rate of depletion as does travel time between patches. Nevertheless, the marginal-value theorem does predict a simple relation between, say average travel time, and average time-in-patch: the longer the travel time, the longer the stay time. The model has also got qualitative support from a few laboratory studies with birds. Foraging birds seem to behave rationally.

Here is another biological example: from the laboratory, where a hypothesis can be rigorously tested, subject by subject, rather than in the field where we can just look to see if it predicts averages. I begin with this example rather than an example of human choice for the reasons I gave at the beginning of Chapter 4: Much better experiments can be done with animals than with humans. Animal experiments can be continued for a long time. Each subject's environment can

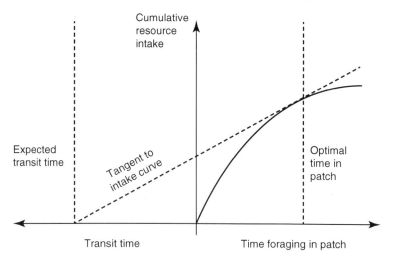

be completely controlled; and the behavior of individuals can be carefully moni-tored—we need not be content with group averages.

As I pointed out in Chapter 3, pigeons, rats, humans, and other animals will continue to make a rewarded response even if reward is only occasional. Skinner and his followers invented many *partial reinforcement schedules* to explore this abil-ity. A particularly effective schedule is the *variable-interval* (VI). A computer tosses a biased coin every second. If the coin comes up heads, a bit is flipped from 0 to 1 and the next response (pecking a disk—called a *key*—for a pigeon, pressing a lever, for a rat) is rewarded (brief access to food for a hungry animal), after which the bit flips back and coin-tossing resumes. The virtue of this procedure is that subjects respond at a steady rate: say 60 or more pecks per minute even if a peck is rewarded (the coin flips 'heads') no more often than once per minute on average. Response rate is sensitive to reward rate: a pigeon may respond at 60 responses/min when rewarded once a minute but, say, 80 responses/min when rewarded twice a minute.

The procedure can be used to study preference. The subject is allowed to choose between two always-available response keys, each dispensing rewards on a variable-interval schedule. Say the two schedules are VI 30 s and VI 60 s (concur-rent VI 30 VI 60): What should the subject do? Well, the one thing he should *not* do is fixate on the richer (VI 30) schedule. The reason is that VI schedules encour-age the animal to *wait* before responding, simply because the chance the coin-flip will come up "1" increases with number of flips (time). So, although the subject might well prefer the richer schedule, his preference should not be exclusive. The longer the time he spends on the richer schedule, the higher the probability that a response to the neglected schedule will be paid off. Just like the patch-foraging birds, a pigeon on the concurrent VI VI procedure can maximize his rate of reward by equalizing marginal utilities, by allocating his pecks so that the marginal payoff rates are the same for each choice.

So much for theory, but the first thing people found out about these schedules, after a few procedural tweaks, is that pigeons precisely *match* the ratio of their rates of response to the corresponding ratio of rewards: $x/y = R(x)/R(y)$, where x and y are response rates, and $R(x)$ and $R(y)$ are the corresponding obtained-reward rates, both averaged over an hour or two[24] (Herrnstein's *matching law*[25]).

Does this mean that the pigeons are 'rational'? Well, yes, in one sense. It is possible to work out how the total payoff to the pigeon, given a fixed budget of pecks-per-minute (he can't respond faster than 80/min, say), depends on how he allocates his pecks between the two choices. The table shows just how total payoff depends on allocation for a three-to-one ratio of payoff. The two VI values are 1 and 3. $x + y$ (the *budget constraint*) must equal 80 responses a minute. The left column shows 4 values for x, the other columns show the corresponding values for y, the two reward rates, $R(x)$ and $R(y)$, and their ratios. The last column shows the total reward rate: $R(x) + R(y)$.

	VI		Values			Total
	1		3			
x	R(x)	y	R(y)	x/y	R(x)/R(y)	R(x)+R(y)
19	0.95	61	2.859	0.311	0.332	**3.809**
20	0.952	60	2.857	0.333	0.333	**3.81**
21	0.955	59	2.855	0.356	0.334	**3.809**
30	0.968	50	2.83	0.6	0.342	**3.798**

The table shows two things. First, the maximum total reinforcement rate is indeed when $x = 20$ and $y = 60$, i.e., perfect matching. But second, deviations from matching don't make much difference. Matching is when $R(x)/R(y) = x/y = 0.333$, but even when $x/y = 0.60$ (bottom row), an 80% deviation from perfect matching, the total reinforcement rate declines by only 3% from the maximum. Only if the pigeon responds very slowly, a response or two per minute, for example, does his preference make much difference to his overall payoff rate. So, it is very unlikely indeed that the matching animal is in some way adjusting its choice ratio until it achieves the maximum payoff rate. Other studies, using more complicated procedures, have confirmed this: if given an opportunity to choose so as to maximize overall (*molar*) payoff, pigeons are unable to do so. If pigeons are 'rational' in these situations, it is emphatically not because they are equating marginals or even comparing the payoffs associated with different response allocations and picking the best. *Yet just this kind of comparison is assumed by most economic accounts.*

From Adam Smith to von Mises, economists have focused on *outcomes* to explain economic behavior. Outcomes do explain—or at least 'are consistent with'—the matching law. But the real explanation lies with von Mises' and Adam Smith's other alternative: not outcome, but *cause*. The pigeons must be doing something, following some response rule, that works well in the VI VI experiment in the sense that it maximizes payoff—even though it is not in any sense *intended* to do so. What that strategy might be, and how it relates to behavioral economics, I will get to in the next chapter.

Two Types of Model

To summarize, there are two kinds of theoretical models: *causal* and *functional*. An example of a causal model is Newton's Second Law of Motion: *force* (F, dependent variable) is directly related to *mass* (*m*) times *acceleration* (*a*) independent variables): Force *F* is the amount needed to accelerate mass *m* by an amount *a*. All the variables can be directly measured, and nothing more is needed to explain the moving body under study.

Most economic models are functional rather than causal. Behavior is explained not by the forces which impel it but by the outcome it achieves. An example of a functional model is the allocation of limited resources between two alternatives so as to maximize subjective value (utility—more on utility in the next chapter). What is the optimal allocation, given that each option is subject to decreasing marginal utility? Suppose the utility of each small increment in commodity A declines in an inverse way: $\Delta U(A) = \alpha/A$; and similarly, for choice B: $\Delta U(B) = \beta/B$, where α and β are positive constants. In other words, the more you have of one option, the less attractive an increment becomes. Given some constraint on the total amount that can be spent, $A + B = constant$ (this is called a *budget constraint*), the amounts of A and B that should be chosen to maximize total utility are then given by equating the two marginal utilities: $\alpha/A = \beta/B$, subject to the budget constraint $A + B = K$, which yields the optimal solution $A/(K-B) = \alpha/\beta$.

Unlike Newton's Law, however, this model is incomplete, since it does not explain how the chooser arrives at the optimal allocation (assuming that he does). He could do it by actually estimating marginal utilities; but, as will see, he could also arrive at the optimum in other ways. This uncertainty about *process* bedevils all functional models. Newton's laws always hold. But functional models only hold under restricted conditions. For this reason, models seem to play a very different role in economics than they do in physics.

Notes

1 *The Trouble with Macroeconomics* (2016) https://paulromer.net/wp-content/uploads/2016/09/WP-Trouble.pdf. Romer is currently Chief Economist and Senior Vice-President at the World Bank.

2 www.theguardian.com/business/2017/jan/01/brexit-slow-burning-fuse-powder-keg-in-2017

3 Rodrik, D. (2016) *Economics rules: The rights and wrongs of the dismal science.* New York: W.W. Norton & Company. Kindle Edition.

4 The comment is by Edwin G. Boring, in his magisterial *A history of experimental psychology* (1957) (2nd ed.). New York: Appleton Century Crofts, p. 642. Philosophers, however continue to be interested in consciousness. See, for example Daniel Dennett's *Consciousness explained*: www.newyorker.com/magazine/2017/03/27/daniel-dennetts-science-of-the-soul. Francis Crick's *The astonishing hypothesis: The scientific search for the soul.* (1995) Scribner reprint edition, and my *The new behaviorism,* (2014) Philadelphia, PA: Psychology Press.

5 Adam Smith (1759) *The Theory of Moral Sentiments.* Section 1.1.24.

6 L. von Mises (1949) *Human action.* https://en.wikipedia.org/wiki/Austrian_School

7 Samuelson, P.A. (1941) The stability of equilibrium: Comparative statics and dynamics. *Econometrica: Journal of the Econometric Society,* pp. 97–120.

8 Hazlitt, H. (1959) *The failure of the new economics.* https://mises.org/sites/default/files/Failure%20of%20the%20New%20Economics_3.pdf.

9 University of Chicago economist Jacob Viner (1933) *Lectures in economics 301* (Price and Distribution Theory).

10 Kaldor, N. (1972) The irrelevance of equilibrium economics. *The Economic Journal,* 82(328), pp. 1237–1255. Kaldor (1908–1986) was a Hungarian-born economist who

explored dynamic approaches to economic problems and was for a while an advisor to the British Labour government.

11 George Soros: Financial Markets. *Financial Times*, October 27, 2009.

12 The leading light was Kenneth J. Arrow: *Collected Papers of Kenneth J. Arrow*. Vols. 1–6. Harvard University Press, 1985. For a survey, see https://en.wikipedia.org/wiki/General_equilibrium_theory

13 Samuelson, P. A. (1955) *Economics: An introductory analysis.* New York: McGraw Hill.

14 *The trouble with macroeconomics* (2016) https://paulromer.net/wp-content/uploads/2016/09/WP-Trouble.pdf

15 Samuelson, P. (1941)

16 $\Delta x = kx(n)[p(n)-g(n-1)]$; $\Delta p = Qp(n)[x(n)-x(n+1)]$, where $x(n)$ is supply at time n, $g(n)$=cost of production at time n, $x(n)$=number produced at time n, and $p(n)$=price at time n. k and, Q are constants (parameters).

17 The point about variability as a topic neglected in favor of unbounded faith in equilibrium is argued at some length in my earlier book: Staddon, J. (2012) *The malign hand of the markets*. New York: McGraw-Hill.

18 https://en.wikipedia.org/wiki/Cobweb_model

19 In B. F. Skinner's sense, meaning the rules relating outcomes to behaviors.

20 www.theatlantic.com/business/archive/2017/01/economism-and-the-minimum-wage/513155/

21 *Noise* is an engineering term for the 'static' that is always added to any signal during transmission. It is assumed to be completely random (*white noise*) or to depart from randomness in a definable way.

22 See, for example, www.salonqp.com/exhibition/info/salonqp-2016-dates-announced/

23 There are other definitions, e.g., rational behavior must be transitive: A > B, B > C implies A > C; and stable: "there is nothing irrational in preferring fish to meat the first time, but there is something irrational in preferring fish to meat in one instant and preferring meat to fish in another, without anything else having changed (Wikipedia)" But of course something *has* changed: the chooser's *history*. Even if rationality = utility maximization, does that mean utility *now* or utility later—over the lifespan, perhaps? Once again, time matters. More on history in Chapter 6.

24 Response rates averaged over an hour or two are termed *molar* rates; as opposed to moment-by-moment responding in real time, which is termed *molecular*.

25 For technical details see Staddon, J. E. R. (2016) *Adaptive behavior and learning*, 2nd edition. Cambridge: Cambridge University Press and also Staddon, J. (2016) *The Englishman: Memoirs of a psychobiologist*. Buckingham: University of Buckingham Press, for a simpler account.

6
BEHAVIORAL ECONOMICS

[T]he premises [of descriptive economics—econometrics—are that] (i) economics is a non-experimental, or … a field experimental, science, and (ii) preferences and technologies are not directly observable or controllable. It follows immediately … that these premises prevent us from answering the most elementary scientific questions.

Vernon Smith[1]

Science does not permit exceptions.

Claude Bernard

Economics is not an experimental science. *Society* cannot be the subject of an experiment, and individual human beings cannot be subjected experimentally to the kinds of strong environmental influences that are unavoidable in the real world. Until relatively recently, therefore, economics was largely descriptive— data gathering and organizing—and theoretical, not experimental. This imposes real limitations.

In the early 1960s, people like Vernon Smith had the idea of running human experiments to model (for example) market equilibrium and the measurement of utility. A few others carried out so-called field experiments, like the 'event studies' discussed in Chapter 7, to try and test models in the real world. At the same time, as I described in the last chapter, economic theory was being applied to experiments with animals.[2] Despite its limitations, the human experimental and theoretical work got much more attention than work with animals.

The most influential human experimental economics is the work of Israeli-American cognitive psychologists Daniel Kahneman and the late Amos Tversky. Their interesting lives, personal interactions, and clever experiments are engagingly recounted by Michael Lewis in a popular book.[3] By the end of 2016,

Kahneman and Tversky's 1979 paper on what they called *prospect theory*,[4] had been cited some 43,000 times, 10,000 more times than *The Origin of Species* (which shows the limitations of citation indices). Before getting to that important paper, I need to describe how human choice differs from choice by animals.

To study animal choice, to get your subjects to choose at all, they must actually experience the different payoffs. They must get occasional food or some other reward for each choice option. The problem looks simpler with people: you can just ask them which option—*prospect*—they prefer. I say 'looks' because as we will see, the superficially simple human situation is much more complicated than the animal one. For example, it is far from certain that the human response to imagined rewards and punishments will be the same as to real ones. Nor can you be certain that the question people are asked is actually the one they are answering. Still, just asking people to choose between notional outcomes does make human experimentation possible, not to mention relatively cheap! Partly because of the difference in experimental methods, the study of human choice behavior has developed along very different lines to the study of choice in animals—although, as we will see, similar principles apply to both.

The properties of human choice behavior revealed in choice experiments with people differ from animal choice in four main ways:

1 People begin with a stock of *wealth*. There is no real equivalent in animal experiments.
2 With no stock of wealth, there is no animal equivalent to *loss*.
3 Individual differences: *There is rarely unanimity in choice experiments with people.* Even when there is a 'statistically significant'[5] preference, 30–40% of subjects may deviate from the majority choice. In most animal choice experiments the effects are reversible, so the result can be replicated with the same individuals, and no statistics are required. In practice, close to 100% of subjects give the same result.
4 Human choice patterns can be changed by experience: risk-aversion can be changed to risk-seeking by appropriate training or prompting, for example. In most animal studies, the pattern of choice under a given reinforcement schedule is stable.

Prospect Theory (PT)

People will usually respond immediately to hypothetical choice questions, such as, 'do you prefer $2,000 with a probability of 0.6 or $1,000 for sure?'—which allows experimenters who use this technique to test hypotheses very quickly. This strategy was adopted by Kahneman and Tversky in a groundbreaking series of papers in the 1970s and succeeding decades. They posed questions to themselves and checked their own answers later with groups of human subjects. They were able to show that people answered their questions in ways that violated the standard version of *utility theory,* the usual outcome-based explanation in economics. The difference between the 'rational' and 'irrational' payoffs in these experiments was rather small. A 'rational' expected value (EV) of, for example, 3,200 versus an 'irrational' EV of 3,000. It seems likely that behavior would have been more rational had the differences been larger, although this variable has not been much explored (the irrational/rational ratio of expected values for most of Kahneman and Tversky's choices was 0.9375).

Swiss mathematician Daniel Bernoulli, discoverer of (among many other things) the principle that allows airplanes to fly, pointed out in the 18th century that the utility of a given increment in wealth is inversely related to the amount of wealth you already have. A hundred dollars is worth much more to a pauper (net wealth: $5) than to a millionaire. Anticipating a now well-known psychological principle we first encountered in Chapter 2, the Weber-Fechner law, Bernoulli proposed that the relationship between wealth and utility is logarithmic, so that equal ratios correspond to equal value. A pay raise of $1000 is worth as much to an employee whose base pay is $10,000 as a raise of $10,000 to one who makes $100,000.

Bernoulli's hypothesis means that the utility function for money is curved— negatively accelerated, *concave,* in the jargon—like the patch-depletion graph I showed earlier. (It need not be logarithmic; many concave curves will do as well.) In other words, Bernoulli's utility function shows that old favorite: diminishing marginal utility.[6] It is the standard, not just for money, but for almost all

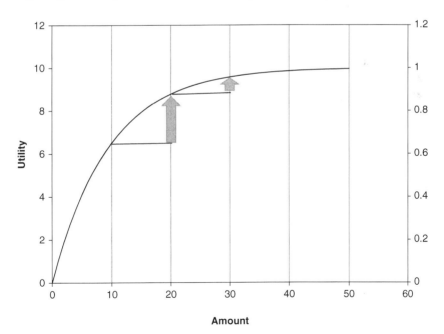

goods.[7] Kahneman and Tversky found many flaws with it, as I will describe. To fix the flaws, they attempted to replace it with their own more complex utility function—with what success we will see in a moment.

The standard concave utility curve predicts that people will tend to be *risk averse* with respect to *gains*. Each increment of money or some other good adds a smaller and smaller increment of utility; hence, doubling the amount of a good less than doubles its utility. Which means that a person should prefer 100% chance of X to > 50% chance of 2X: $100 with probability 1.0 over $200 with probability, say, 0.6.

Kahneman and Tversky found this effect in one of their experiments. In Problem 3 of the prospect-theory paper, 95 people were asked to choose between two outcomes: 4,000 with probability 0.8, versus 3,000 for sure. 80% (note: not 100%) chose the sure thing, even though it is less than the *expected value*[8] of the gamble: 3,000 < 3,200. Apparently, for most people 0.8 times the utility of 4,000 is less than the utility of 3,000, 0.8xU(4,000) < U(3,000)—because the utility function is concave.

Yet expected value alone explains Kahneman and Tversky's Problem 4, where 65% of the same 95 people preferred a 0.2 chance of 4000 to a 0.25 chance of 3,000: in expected values, 800 was now preferred to 750—no sign of diminishing returns there. Evidently *certainty* has an effect that goes beyond the probability of one.

There is another way to look at Problem 3. To deal with the special effect of certainty, that it seems to have an effect beyond simply being a probability of one, Kahneman and Tversky added an assumption: that the gamble—$3,000 for sure versus 0.8 probability of $4,000—actually involves a perceived possible *loss* since the '3,000 for sure' option is guaranteed. It increments your current state of wealth, *unless* you choose the risky option, in which case, with probability 0.2 (the chance you *don't* win the 4,000) your wealth could go down by 3,000, a *loss*. So the issue becomes: is the absolute value of 0.2xU(−3000) greater than the absolute value of U(4,000–3,000). That is, is 0.2 times the cost of losing 3,000 (i.e., actual expected value 600) greater than the benefit of a net gain of 1,000. And the answer is 'yes' if the utility curve is rotated through 180° into the lower-left quadrant. In this case the disutility of a loss of 600 is greater than the utility of a gain of 1,000. The same curvature that predicted a preference for the fixed option—risk aversion—predicts a preference for risky option—risk seeking—when the choice is between losses.

The shift of origin is not really required for Problem 3, since the all-positive utility curve also works. But it *is* necessary to account for some other results.

Notice that shifting the origin of the utility curve for Problem 3 requires an assumption about how people *perceive* a choice situation. Reference level is always set by the subject's expectation/perception, as the following memorable exchange illustrates. The scene is investment banking house Drexel, Burnham,

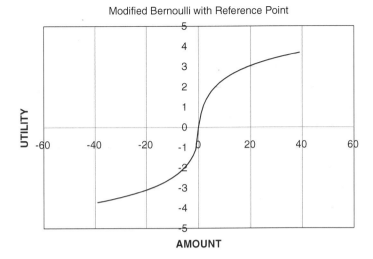

Modified Bernoulli with Reference Point

Lambert (defunct since 1990) in 1985. A young banker is brought in to learn his bonus for the year:

> At their session, Kay reviewed Levine's work during the year, telling Levine he felt he "could be relied upon to take a transaction and work on it at the highest level of confidence." Then, feeling certain he was about to bestow on Levine a sum that exceeded even Levine's measures of his own abilities, Kay said, "Your bonus for 1985 is … one million dollars." [which was a lot of money in 1985]
>
> "That," Levine replied, "is an insult." He stood up and stalked out of Kay's office[9]

The subject sets his own reference level.

Their data led Kahneman and Tversky to propose what they called *prospect theory* as a replacement for standard utility theory. But introducing the idea of perception adds a complication. It means that prospect theory is a *hybrid* of process and outcome, of causation and teleology. The 'front end' is process—what Kahneman and Tversky term *editing*: "Prospect theory distinguishes two phases in the choice process: an early phase of editing and a subsequent phase of evaluation."[10] Editing is their name for the set of cognitive processes that set up the problem for the *evaluation* part of the theory.

Editing is a process, but it is not defined in any calculable way. Terms such as framing (added in a later version[11]), combination, segregation, isolation, coalescing, and cancellation are names for operations that the decision maker can use to simplify and structure the choice problem so that it can be evaluated. In fact, framing, cancellation, etc. are simply *labels* for the variety of ways Kahneman and

Tversky's experimental results deviated from the predictions of expected utility theory. These terms are silent on the actual underlying processes, on exactly when these processes are invoked and precisely how they work. Hence, a large chunk of the most influential theory in behavioral economics is in large part just systematic folk psychology.

The well-defined part of prospect theory is the second phase, *evaluation*, which is a modified version of expected utility. Evaluation is a teleological/functional theory; it describes an outcome but, again, not the process the system uses to achieve it.

Evaluation, according to PT, modifies standard utility theory in three ways:

1. We have already encountered the idea of a *reference point* or adaptation level. The reference point moves the origin of the utility curve to a point representing the subject's perception of his current wealth. A choice option that yields a state of wealth less than the reference point is perceived as a loss. Adaptation level is not an implausible assumption: the idea that people are sensitive to changes rather than absolute values is a core principle of perceptual psychology.[12]

2. The reference-point assumption means that the upper–right quadrant of the utility graph is the same as before. But the lower left part is modified, as in the previous picture, to deal with the following three results involving losses. The last two require changes in the standard form.

First, most people will choose a 50% chance of losing $100 only if the other possibility is a win of $200 or more: "people tend to be risk averse in the domain of gains and risk seeking in the domain of losses."[13] Thus, the first part of the south-west quadrant must be curved downwards, like the previous graph.

But, second, 65% of subjects preferred a 0.2 chance of winning 4,000 over a 0.25 chance to win 3,000 (expected values: 800 > 750; Kahneman and Tversky Problem 4), i.e., they were in line with expected-value theory and not risk averse. So, the curve must straighten out at intermediate probabilities. And finally, 86% of people preferred a 0.9 chance to win 3,000 over a 0.45 chance to win 6,000 (Problem 7), even though both gambles have the same expected value. Now subjects showed risk aversion. Hence the curve in the SW quadrant must begin steep, then straighten, and then flatten out. This is the iconic prospect theory utility graph (Kahneman and Tversky, Figure 3), which resembles the rotated Bernoulli in the previous picture.

3. Gambles at extreme odds do not fit even this modified utility graph. For example, 73% of people prefer a 0.001 chance to win 6000 over a 0.002 chance to win 3,000 (Kahneman and Tversky, Problem 8), even though the expected values are the same, i.e., back to risk-seeking. Another exception is provided by something called probabilistic insurance. Suppose that you are indifferent about whether or not to insure your house against earthquakes. A creative salesman then makes you this offer: we are willing to insure you for just 45% of the original premium if we are allowed to toss a coin after an incident to decide whether we pay you or just refund your premium. Obviously, this is a better deal than the original

one, which you were on the point of accepting. Yet people will usually reject it. Not that they are against probabilistic insurance on principle. As Kahneman and Tversky point out, the decision to quit a risky habit like smoking is a form of probabilistic insurance.[14]

These problems cannot be solved through changes in the utility curve alone. To fix them, Kahneman and Tversky re-introduced a familiar strategy:[15] a somewhat untidy transformation of probabilities ("π is not well-behaved near the endpoints."), termed *decision weights*, that parallels Bernoulli's transformation of value (Kahneman and Tversky, Figure 4). Evidently both the utility curve and the probabilities must be transformed to accommodate data.

A scientific theory usually consists of a set of assumptions and a set of predictions deduced from them. The number of assumptions must be smaller than the number of predictions if the theory is to be useful.[16] Prospect theory, in the 1979 paper as well as later revisions, is not a real theory in this sense at all. It lists many surprising effects, gives them names such as endowment effect, anchoring, and so on and then provides an incomplete, hybrid account of the whole process. It is basically a catalog—orderly and eye-opening, to be sure, but a catalog nonetheless—a catalog of exceptions to standard utility theory. The standard, all-positive, concave utility function and marginalist mathematics fails in many ways. Yet the prospects for prospect theory as a viable alternative are mixed. Is there some way to go beyond PT to a real theory?

Behavioral Economics: Animal Division

First, I think we can rule out making a formal model based on PT. Attempts have been made. The models that have been proposed are too complicated to be useful. The great computer pioneer, physicist, game theorist and bon vivant, Hungarian-American John von Neumann famously wrote: "With four parameters I can fit an elephant, and with five I can make him wiggle his trunk." Attempts to formalize prospect theory seem to need four or five parameters.[17]

Human choice behavior is affected by many things; not just the numbers you are presented with, but your state of wealth, your experience, both recent and historical, the kind of goods on offer—money or things, for example—the context in which the task is presented, the way questions are asked and your willingness to think through the options. These factors affect choice in ways that pose problems for any theory.

In his 2011 book, *Thinking Fast and Slow*, Kahneman cuts through this complexity by proposing that prospect theory only deals with decisions that are made quickly. He distinguishes between *fast* and *slow* cognitive processing, what he calls System 1 (fast, thoughtless) and System 2 (slow, reflective). Presumably, the rapid process is simpler to understand than the slow one. Kahneman, an experimental psychologist not an economist by training, has evidently begun to shift his attention away from utility/consequence-based thinking to the actual processes—the actual *causes*, in the language of Adam Smith and von Mises—that underlie choice behavior.

Causes have long been the focus of animal-choice research. In fact, the fast-slow distinction is indistinguishable from a familiar dichotomy from the animal lab. Another critical prospect-theory concept, *framing*, also has an animal-lab counterpart.

So, let's look at four aspects of prospect theory from a causal point of view, the point of view of choice research with animals. The four aspects are:

1   the role of consciousness;
2   framing;
3   the fast-slow distinction;
4   individual differences.

1   Consciousness

Although there is some debate about the issue, most would agree that consciousness is a very human attribute. If consciousness is involved in human choice, as many cognitive analyses assume, animal-choice research may seem less relevant. Prospect theory is usually considered a cognitive, as opposed to a behavioristic, system, yet the role of consciousness, especially in the 'fast' system is minimal: "The mental work that produces impressions, intuitions, and many decisions goes on in silence in our mind,"[18] and "Studies of priming effects have yielded discoveries that threaten our self-image as conscious and autonomous authors of our judgments and our choices."[19] says Kahneman. This is really no different from the standard behaviorist (usually opposed to cognitivist) position. For most behaviorists, most brain activity, most thought, is completely unconscious. The autonomy of the unconscious isn't even a very new idea. Well before Sigmund Freud, the eccentric genius Samuel Butler (1835–1902) referred frequently to the unconscious in his posthumously published novel *The Way of All Flesh*, which is in effect an autobiographical study of his own psychological development. Darwin's cousin, Francis Galton (see Chapter 4) made similar observations[20] and author C. P. Snow (famous for his 'two cultures' distinction between science and the humanities) spoke of an unconscious 'secret planner.'

Cognitive psychologists are admittedly more interested than behaviorists in consciousness. But the contemporary cognitive view, like Butler's, is that consciousness is not an active agent, but rather something like a 'workspace' which "allows the novel combination of material."[21] Cognitivists like Kahneman no longer differ greatly from behaviorists in this respect. For both, human choice behavior, especially if a decision must be made quickly, is largely determined by unconscious processes.

2 *Framing*

Framing is the term Kahneman and Tversky give to the effect on a choice of context and the way a problem is presented. The term does not occur in the 1979 paper but, with a number of other labels such as nonlinear preference and source-dependence, was made necessary by new data that did not fit the original formulation. An example of framing is:

> The statement that 'the odds of survival one month after surgery are 90%' is more reassuring than the equivalent statement that 'mortality within one month of surgery is 10%.' Similarly, cold cuts described as '90% fat-free' are more attractive than when they are described as '10% fat.' The equivalence of the alternative formulations is transparent, but an individual normally sees only one formulation, and what she sees is all there is.[22]

The way in which a question is asked (this is sometimes rather pretentiously termed *choice architecture*) can have a large effect on the subject's response.

Framing is just another name for the effect of a subject's *history*, his past experience, on his expectations (equivalent to his response *repertoire*, as I will show in a moment) at the time he makes a response. The history can be brief—like the question itself, the way it is posed—or it can involve earlier events. Pavlovian conditioning, which I discuss next, is an example of a lengthy history that has a framing effect.

Pavlovian (also called *classical*) conditioning is one of two processes involved in trial-and-error learning (also called instrumental, trial-and-error, or *operant* learning). The processes are the ones that Charles Darwin identified as the means by which species adapt and evolve: *variation* and *selection*. In evolution, the variation is provided by genetic mutation and recombination (among other things). The selection is *natural selection*, the reproductive advantage of some types over others. In learning, the variation is just the *repertoire* of potential activities that an animal, with a given history and in a given situation, is capable of. The repertoire is largely set by classical conditioning. Repertoire can be thought of as a measure of the organism's (largely unconscious) *expectation*. The selection is provided by the consequences of action: reward or punishment—operant conditioning. Successful—rewarded—activities are strengthened; unsuccessful, or punished, activities are weakened. Notice that this variation-selection approach allows for individual differences. People may begin with different repertoires. Individual differences are not easily accommodated by rational-choice theory.

An organism's repertoire is determined by expectation which is determined by the *context*—the prospect of food or threat or a mate, for example. Context includes the actual stimulus environment and the associated history (the organism's experience in that environment or similar environments) up until that point. Stimulus context is like a spotlight playing over a nighttime crowd of demonstrators in a public square. It shifts from place to place, from context to context. At each position, a different group is illuminated and different placards can be seen, one (the active response) higher than the others (silent responses). Demonstrators move through the crowd, sometimes in the spotlight, sometimes not. But in each interval of time, only a small section of the crowd is visible, one individual stands out, and the repertoire is restricted.

A repertoire is limited, less so in humans than in animals, no doubt, but limited in all. For example, in the presence of a stimulus—a colored light, or even a brief time after a time marker—that has signaled a high probability of food in the past, a hungry pigeon will show a very restricted repertoire. The hungry bird will peck the light, even if pecks turn it off and he gets no food on that occasion—a very 'irrational' behavior. A hungry raccoon trained to pick up a coin and deposit it in a piggy bank for food reward will learn to do it. But then, having learned the association between coin and food, it will start 'washing' in its characteristic way, so preventing the delivery of more food. My dog, when I begin to put on my shoes, signaling an imminent walk, jumps up and impedes me, delaying a much-desired outcome. All 'irrational,' all the result of limited behavioral variation induced by a 'hot' situation.[23]

These maladaptive behaviors are examples of what has been called *instinctive drift*. The procedures that gave rise to them are all examples of Pavlovian conditioning, the formation of an expectation by the animal based on a past association between a stimulus and reward. This expectation in turn defines the set of activities—the repertoire—that, depending on species and individual history, the animal is capable of showing in this situation. If the pigeon and the raccoon expect imminent food, they engage in food-related activity. If they expect to be attacked or to encounter a potential mate, they shift their expectations, and their behavioral repertoire, accordingly. This process of "Pavlovian framing" is easy to demonstrate experimentally. Framing in humans is less dramatic, but no different in essence.

Note that the existence of framing means that reward is not reward. Rewards are not all the same—as most economists seem to assume. Money is not the same as love. The repertoire that is induced by the expectation of cash reward will be different from the repertoire induced by love of ideas or a wish to cure patients. Consequently, paying teachers or doctors more will not necessarily produce better teaching or medical care. And old habits—links between a situation and a repertoire—are not immediately erased by new circumstances. Habits persist: economists should not be puzzled by the fact that people tip in a restaurant they expect to visit only once.

The role of framing—expectation—in human choice can be tested experimentally. Humans are neither as controllable nor as dumb as pigeons. But the facts that '90% fat-free' produces a more positive expectation than '10% fat,' and that we are more inclined to go for surgery if we are told that the probability of success is 90% than if we are told the probability of failure is 10%, shows that the same process is at work. The now-popular 'nudge' approach to social engineering is another example. When required to 'opt-in,' for example, to a company health-insurance plan, people hesitate. They hesitate to do anything that is not either habitual or obviously rewarding. So, they will also hesitate to 'opt-out' if the plan is in place and they must respond to end it. If we want them to have a plan, 'opt-out' with a plan in place if they do nothing is the obvious way to ensure the desired result.

Pavlovian effects are common in human behavior. Notoriously, even extremely rare pairings of a particular kind of situation and a very bad outcome can have large effects. Air travel went down massively after 9/11. But air travel also declined after lesser, terrestrial terrorist events even after their rarity had become pretty well established. Tiny penknives were not allowed in aircraft cabins, even after locked pilot-cabin doors became mandatory and made the exclusion irrelevant to the survival of the plane—because box-cutters were used on 9/11. Resistance to e-cigarettes from the health-and-safety community is strong, despite lack of evidence of their ill effects—because of the pairing of smoking and disease. Sales of the medicine Tylenol went down by as much as 90% in 2010 after contamination was found in a few batches. The Pavlovian pairing of a bad-news story—never mind the details, still less the actual statistics—with the name of a popular product often has disproportionately large effects on sales.

Perhaps the quickest way to show how a change in the stimulus situation changes the repertoire of human subjects, and elevates one response over others as the 'active' response, is something called *priming*.[24] Priming is an example of framing. The terms are different, but framing and priming are both effects of environment on expectation—on repertoire. Priming works like this. The subject is asked to fill in the blank letter in, for example, FL_T. After first seeing a picture of a *butterfly,* she is likely to say FLIT. But if she first sees a picture of a *table,* she is more likely to say FLAT. *Butterfly* at once evokes expectations about butterflies—they fly, flutter, have bright colors, etc. *Tables,* on the other hand, are square, flat, have legs, etc.

Another example of framing is something called *confirmation bias*, which is just a fancy name for the fact that most people more readily notice facts that support an existing belief than facts that are contrary to it. Again, there are exceptions. A good inductive scientist, probably as a consequence of long training contrary to instinct, will be at least as sensitive to discrepant as to supporting evidence. Framing, priming, and confirmation bias show how context creates an expectation that affects both the available repertoire and its strongest member, the active response.

The take-home is that a subject's response to a question is limited by his repertoire. Repertoire is limited by history—the instructions he has been given, the examples he has seen, the expectations he formed at the beginning of the experiment. Given the right instructions and examples, there is little doubt that all of Kahneman and Tversky's subjects would have behaved rationally—or, with a different history—all would have behaved irrationally. The available repertoire, the state of the subject when the question is asked determines his response. The way that state is determined—by instructions, examples, and events in the subject's past—is the real key to understanding his behavior. Without knowing how history sets context, how a person's past experience determines his repertoire, data like the apparently 'irrational' behavior of Kahneman and Tversky's subjects, raises more questions than it answers.

3 Fast-Slow or Active-Silent?

Let's look again at Kahneman's fast-slow distinction. He contrasts the quick answers he and Tversky got to their choice questions—answers which were often 'irrational' in the sense that they did not maximize gain—with the slower and more 'rational' answer that most people arrive at after deliberation. The quick system responds with answers that are more 'available' or 'accessible' than other responses which may in fact be better. Kahneman also tells his readers to "remember that the two systems do not really exist in the brain or anywhere else." They are ways to talk about the fact that people may respond one way at first and another way, given some time for reflection.

'Fast' and 'slow' just mean 'accessible' and (relatively) 'inaccessible.' Exactly the same distinction has emerged from the study of choice behavior in animals. 'Accessible' behavior is just the first response to occur in a new situation: the *active* response. Other—*silent*—responses, the rest of the repertoire, qualify as 'inaccessible.' If the active response is rewarded, it is strengthened—*selected*. If not, it is weakened until it is supplanted by the next-strongest, silent response. But if the repertoire is restricted, an active response may persist even if it is ineffective, as in the 'instinctive drift' examples I gave earlier. Just as people may respond in ways that defy expected-utility theory, animals also can respond in ways that reduce rate of reward.

How is a response *selected* from the available repertoire? In other words, how does the process of *reinforcement* actually work? Again, animals are easier to understand. I begin with the simplest possible choice situation. The two-armed bandit is popular not just in Las Vegas,[25] but among operant conditioners and those who study learned instrumental behavior in animals. It involves two simultaneously available choice options, each paying off randomly, with different probabilities. Pigeons and rats often seem to follow elementary economics. They are rational, in the sense that they maximize the rate of reward, although they don't do it by explicitly comparing long-term (molar) averages, as I showed earlier in connection with the so-called matching law. But in simple situations they really *seem* to maximize. For example, a hungry pigeon in a Skinner box randomly paid off on average for one in every ten pecks (termed a *random ratio ten* schedule) on a response key on the left and on average for one in every five pecks (RR 5) for pecking on the right—such a pigeon will soon peck only on the right. It *maximizes*.

But how does he do it? Is the pigeon really comparing the two options and then picking the best, as most economists would assume? There are hints that the bird is not as smart as he looks. Sticking with ratio schedules, consider a *fixed*-ratio schedule that requires exactly 100 pecks for each food reward. A pigeon will soon learn to get his food this way. But he will always *wait* a second or two after each reward before beginning to peck, thus delaying all future food deliveries unnecessarily. Fatigue? No; on a comparable *random* ratio, where rewards occasionally occur back-to-back although on average only one in 100 pecks will pay off, he will respond steadily, not waiting after food.

The reason he waits on the fixed ratio is that he has a built-in, automatic timing mechanism—a *heuristic*, if you like—that responds to the minimum time between food deliveries enforced by the fixed- (but not the random-) ratio. The food-food delay enforced by the time it takes to make 100 pecks causes an automatic pause after food. It's easy to prove this by comparing two kinds of *time-based* schedules. Suppose you start a 60-s timer after each food delivery and don't reward a peck until 60 s has passed (this is called a *fixed-interval* schedule). As you might expect, once the animal has learned, he waits perhaps 30 s after each reward before beginning to peck. If his time sense were perfect, no variability, he would presumably wait exactly 60 s so as not to waste pecks. His timing is a bit variable so he usually starts early. Perfectly rational: don't waste pecks, but don't delay food unnecessarily.

Now suppose you modify the procedure slightly by only starting the 60-s timer after the first peck. The rational adaptation is simple: respond immediately after food, and then wait as before. But pigeons don't do this. If the interval duration is T s and the typical wait time is $T/2$, then on this response-initiated-delay schedule they will wait not $T/2$ but T s, delaying reward quite unnecessarily. The bird is still telling time from the last reward, not the first peck after reward. The reason is that the bird's memory system has limitations. A pigeon usually times from the last event he can remember that predicts the time-to-food. He should therefore time from the first post-food peck. But he can remember the time of the last food much better than the time of the first post-food peck, which is just one peck among many.

The point is that rational-choice theory is of limited use in understanding what these animals are doing. Animals never, and humans almost never, are *explicit maximizers*, computing marginals and choosing 'rationally.' They adapt via built-in processes, like timing and instinctive drift, that sometimes yield 'rational' behavior and sometimes don't. They select from a limited repertoire which is made up of what behavioral ecologists call 'rules of thumb' or (a familiar term to decision theorists) heuristics. German decision theorist Gerd Gigerenzer refers to the contents of the repertoire as an *adaptive toolbox*. Artificial intelligence pioneer Marvin Minsky labeled the multitude *society of mind*.[26]

What Is Rational?

Before I say something about the kind of process that might underlie the behavior of a choosing pigeon, it is worth looking a bit more closely at what the idea of rational behavior really entails. Is the pigeon's behavior *really* irrational? Is it really irrational to wait a second or two before beginning a run of 100 pecks, but not to wait after food on a random ratio, where the next food might require from one to 200 pecks? The answer is: it depends on the most important, and most frequently evaded, ingredient in the concept of rationality as usually discussed. That is the fact that any claim that behavior is, or is not, rational is essentially meaningless unless the *constraints* under which the system is operating are fully defined. There are two kinds of constraint: external—in finance, a limit on budget, in animal studies a limit on time. External constraints are acknowledged by economists. But there are also *internal* constraints, limits on memory, computational capacity, rate of activity, etc. Without knowing the internal constraints, the optimizing economist places himself in the position of an all-know-all-seeing Deity—or at least Carnak the Magnificent. Knowing all that is to come, the economist now pronounces on the very best behavior and, sometimes, finds his subject wanting. But what if he is far from all-knowing, but is in fact willfully blinded, looking only at a fatally over-simplified version of the problem under study?

Rational behavior, to an economist, is behavior that maximizes something, usually something monetary. So economic rationality is an example of what is called an *optimization problem*. The online *Britannica* says this about optimization problems:

> Optimization problems typically have three fundamental elements. The first is a single numerical quantity, or *objective function*, that is to be maximized or minimized. The objective may be the expected return on a stock portfolio, a company's production costs or profits...The second element is a collection of variables, which are quantities whose values can be manipulated in order to optimize the objective. Examples include the quantities of stock to be bought or sold, the amounts of various resources to be allocated to different production activities...The third element of an optimization problem is a set of *constraints*, which are restrictions on the values that the variables can take.[27]
>
> *[my emphasis]*

Economists almost invariably have money as the first element (objective function), and some numerical variables as the second element. But apart from budget constraints (external) they usually ignore the third element, the internal constraints on the optimizer. (I will have more to say about economics when we get to the efficient market hypothesis, in Chapter 7.) But the constraint element is absolutely vital to understanding optimality/adaptiveness, as I can show by example.

Look again at the pigeon in a response-initiated fixed-interval (RIFI) schedule. The 'rational' strategy is: respond right after food, to start the timer, then wait for half (say) of the interval before beginning to respond. But think about just what is required by way of memory to *learn* this strategy. The pigeon must be able to remember the time of occurrence of his first post-food peck in each interval, call it T_0, so that he can learn that time-to-the-next-food is fixed with respect to T_0 but not with respect to the times of subsequent pecks or time since the last reward, T_1. He must also learn that T_1 is always longer than T_0, so that he should use T_0 and not T_1 as the time marker. In other words, the lowly bird must have a very good memory, a memory with temporal resolution sufficient to identify the time of occurrence of the first post-food peck and represent it separately from later pecks.[28]

How useful would such a great memory be to pigeons? How likely is it that pigeon evolution would have developed it? The answer to the first question is pretty clear. Situations like the RIFI schedule are rare to nonexistent in the evolutionary history of pigeons. So, the high-res memory necessary to adapt optimally to them is also unlikely to have evolved. That is a mechanistic explanation for the pigeon's failure. But there is also a maximization-type explanation. An excellent memory involves investment in the neural tissue that must underlie it. This involves resources and tradeoffs with other abilities—foraging, spatial memory, etc.—that may be much more useful to the pigeon species. In short, once we understand the constraints to which real pigeons are subject, our bird's failure to behave 'rationally' in the RIFI situation doesn't look irrational at all. A simple memory, which causes unnecessary waits at the beginning of a long fixed-ratios and on RIFI schedules, may nevertheless be worth it in terms of the overall 'fitness'[29] of the pigeons. It may not be rational for the observing economist, but it is rational for the pigeon species.

A Causal Model

So, what *are* pigeons doing in these simple choice situations? The time dimension adds complexity and the way in which time is incorporated into learned behavior is still not perfectly understood. So, let's stick with random-ratio schedules and the two-armed bandit choice situation, where time is irrelevant. The process by which a pigeon chooses between two such schedules seems to be surprisingly simple.[30] The rules are:

1   Response tendencies (strengths)—peck left or peck right—compete.
2   Response strength of response A is proportional to total cumulated total reward for A divided by total cumulated A responses (i.e., a Bayesian payoff history), and similarly for B.
3   Response rule is: Strongest response wins, becomes the *active* response.
4   Reinforcement process: active response just rewarded? Strength increases. Not rewarded? Strength decreases, according to 2. above.[31] *Silent* responses: no change in strength.

Crude as it is, this version of the process makes some predictions that cannot be matched by a utility-type theory. A good test-bed is a choice situation where

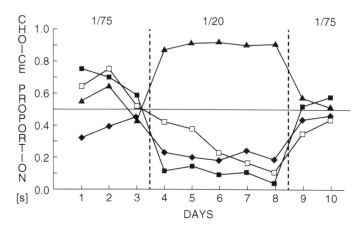

there is no rational strategy or, to put it a little differently, *all* strategies are rational. Such a situation is choice between two *identical* random ratios. It makes no difference to overall payoff how the animal allocates his responding between the two schedules. So, no rational-choice theory makes any prediction. Since this is a situation where the subject actually experiences the outcomes rather than being interrogated, prospect theory is not relevant. But the selection–variation process I have described does make a straightforward prediction. If both random ratios are small, for example 20, the pigeon should settle on one choice exclusively; but if they are large, for example 75, he should be indifferent.

The (admittedly limited) data available do indeed support this prediction.[32] The picture shows the preferences of four individual pigeons over a ten-day period when they were exposed each day for an hour or so to the two random ratios in ABA sequence: 75:75, 20:20, and 75:75. As you can see: the richer the schedule, the more likely the birds were to fixate on one side or the other, even though both choices yielded the same payoff. These are old data, whose significance was not clear at the time. But now it seems obvious what they are telling us about how the animals choose between apparently identical options. The pigeons are not optimizing in any explicit or intentional way. The rewarded response is being strengthened and the unrewarded one weakened; that's it. The Darwinian selection–variation approach to choice seems to work quite well. It has the advantage of being a causal rather than a teleological process, so can easily accommodate deviations from 'rationality'—as well as individual differences.

4   *Individual Differences*

Individual differences are almost completely ignored in choice experiments with humans. Indeed, they are largely ignored in much of NHST (null hypothesis significance testing) social psychology. Yet they cannot be neglected if the aim of the research is to understand the behavior of individual human beings rather than aggregates. As I pointed out in connection with learning curves in Chapter 2, group data are fine for marketing and polling, surveys and voting, but grossly inadequate if the aim is to understand how individual human beings work.

Kahneman and Tversky's subjects showed large individual differences. Groups were never unanimous in their choices—which is surprising given the precision and relative simplicity of the problems with which they were confronted. The majorities were 'significant' statistically although, as you may remember from the earlier discussion (Chapter 3), the standards for significance that were acceptable in 1979 are now suspect. Although probably replicable, Kahneman and Tversky's results are nevertheless quite variable: as few as 65% of subjects favored the discrepant, utility-theory-contradicting option in some experiments.

What are the discrepant subjects doing? Indeed, what are the majority doing? Are they all in fact doing the same thing? The prospect-theory tasks are so simple we can learn little from looking at the subjects or even asking them—and introspective reports are in any case notoriously unreliable guides to what is actually going on. The springs of rapid choice are mostly unconscious. Nevertheless, it would be helpful to get some idea of what subjects expected at the beginning of the experiment. It would also be interesting to ask the question in different ways: What is *your* choice? What would most people choose? What is the optimal choice? In other words, are subjects understanding the task in the way the experimenters intended? Or are the questions ambiguous in some way, so that different subjects are actually doing different things? These problems of interpretation do not exist in animal studies, where the subjects actually experience outcomes and the procedure by which they obtain them—rather than having to interpret verbal instructions.

It is also worth asking an obvious question. The subjects in these experiments—and many other like experiments[33]—seem to be so dumb. How come the experimenters are so much smarter than their subjects? Why don't *they* show all the illogicalities and rigidities they have successfully demonstrated in their subjects?

Are the experimenters that much smarter? Well, probably not, or at least not in every case. The difference, of course in their history and the repertoire they bring to the situation. The subjects have little history, or a history slanted (by the experimenters) in one direction or another. The experimenters have spent months or years trying to come up with situations that will (not to mince words!) trick their subjects into saying something silly. The subjects' responses have little consequence. They are given no opportunity to think at length about their answers, or to consult with others. It is no wonder that many of them respond in ways that, to dispassionate outside observers, may look biased or dumb.

The Effect of History

Unraveling causation in experiments with human beings is very difficult. The PT tasks, which are all one-off, don't allow the experimenter to do what is necessary.

Subjects are not repeatedly asked the same question, shown the actual outcome, and asked again. There is no exploration or manipulation of the subject's history. On the other hand, the pigeon choice experiment is a sort of 'animal model' for the human problem. It is an individual-subject task that allows us to separate outcome/utility-based theories from a well-defined process. The aim is always to understand the repertoire which the subject brings to the situation which is then winnowed by consequences.

In the pigeon experiment, the repertoire was simply peck left or right. The repertoire in the prospect theory problems is impossible to measure directly. Nevertheless, we can get an idea of the responses/heuristics available to subjects by looking at the effects of *pre-exposure*—giving subjects different experiences before asking them the problem question. I have already described the effects of priming and framing in general. There is little doubt that giving subjects a little lesson in probability theory would reverse at least some of Kahneman and Tversky's results, which depend rather critically on the relative naiveté of the subjects. Repeating the question, allowing an outcome—they get one or other option—and repeating again and again, would certainly shift most subjects away from risk-aversion to 'rational' choice. A series of AB-type *transfer experiments*—presenting hypothetical results from other experiments, some 'rational' some not as treatment A, and B as the Kahneman and Tversky procedure—might change subjects' B choices. These changes, in turn, might provide clues to the repertoire available to them in PT-type experiments and help explain both individual differences and common 'errors.'

Kahneman and Tversky give some hints about possible underlying processes in their discussions of various 'effects.' Three examples are the *endowment effect, representativeness,* and *anchoring.* The endowment effect, a form of *loss aversion,* is the fact that people seem to add value to something they already own. For example, you buy a decorated mug on sale for $10, which you declined to buy pre-sale for $12. But now, you refuse to sell it to a friend for $15. Is this a built-in bias, or does it depend on certain kinds of experience?

As for representativeness, consider the following question; what is Jean's profession, given that 30-year-old Jean is literate, active in social causes, has short hair and is unmarried? You are given eight choices, including radical journalist and accountant. Most people will probably pick the journalist, even though there are many more accountants than radical journalists—because the thumbnail about Jean is a closer fit to a 'typical' radical journalist than a 'typical' accountant. When answering questions like this, most people ignore what is called *base rate,* in this case the relative numbers of journalists and accountants in the population: few journalists, many accountants, which means that Jean is much more likely to be an accountant than a journalist.

Representativeness is an example of what animal psychologists called *stimulus generalization* or *similarity judgment.* The subject is asked to guess which class the subject—Jean—falls into. Instead, he guesses which category she most resembles.

The problem posed is two-dimensional—it involves both base rate and similarity. The subject judges based on similarity alone. Given some earlier two-dimensional problems, would the subjects have done better on this one? Told that the task was to identify her profession not which description seems most like her, would subjects improve?

Anchoring, like priming, is another example of framing. People are asked to make a quantitative judgment in an area about which they know little or nothing—like the number of African nations in the United Nations. But first they are shown a random (wheel-of-fortune) number between, say zero and a hundred. If the number was close to 100, they guess high, if close to 0, they guess low—even though they know the number should be irrelevant. The random number they saw just before being asked the question obviously 'primed' them so that numbers in that vicinity were momentarily stronger than the rest.

Kahneman and Tversky designed questions that clearly misled many unprepared subjects. If, instead, they had chosen to give all their subjects a set of experiences designed to control their expectations about the task, there is little doubt the variability of their choices would have been reduced. Their choices might well become both unanimous and, depending on the experiences, more 'rational' also. Explorations along these lines could tell us much more about the way repertoires are determined. We might learn more about what subjects are really doing, rather than just what they are *not* doing: i.e., being 'rational.' We might learn something about the evolution of repertoires and hence more about both 'rational' and 'irrational' behavior.[34] We might learn something about how to educate.

Logic

I earlier discussed the concept of intelligence, presumed by many to be a single measurable entity like, weight or eye color. It is not so simple of course. Intelligent behavior depends on many mental/brain processes whose details are largely unknown. A similar notion in economics is *rationality*, which is also seen as a unitary, unanalyzable guide to action. The various biases and effects that have been identified—representativeness, priming and so forth—are usually portrayed as alternatives to rational behavior. But the selection/variation approach to learning shows that rational behavior is itself just one product of processes that sometimes seem to misfire—at least from the point of view of rational choice. But rationality is not some perfect Platonic universal that stands in opposition to fallible rules-of-thumb. *Au contraire*, it is just the product of the selection/variation process one of whose components—and by no means the most useful one—is the operations of logic.

Here is an example of apparently illogical behavior. Kahneman and Tversky showed that *under some conditions* people make an obvious logical error when asked certain kinds of question: first, they are presented with a description of a lady, *Linda*. The description mentions various things about Linda: she is outspoken, concerned about social issues, likes numbers, and so forth. Then subjects are asked to pick the most likely of several possibilities, two of which are:

1   Linda is a bank teller.
2   Linda is a bank teller and is active in the feminist movement.

When the two options are juxtaposed like this, it is obvious that 1 is more likely than 2, because 1 contains 2. No one should pick 2 over 1 and very few will do so. But when presented as Kahneman and Tversky did, not directly compared, but embedded in a list of distractor questions, the preference was reversed. One group of subjects saw four possibilities, one of which was 1. The other group saw the four possibilities but 2, was in place of 1. More chose 2 than 1, an obvious logical error.

But is this a failure of logic, or a trick question which seems to ask the subjects to find not which class does Linda fall in, but how many attributes does she share with the description? Obviously, Kahneman and Tversky's subjects were not doing logic. But were they irrational—just stupid? Probably not; probably just interpreting the question in a different way. The 'errors' they made are not errors if the question they heard was the second one.

The point is that people can do many things: judge representativeness/similarity, respond to hints (priming), and context (framing)—as well as logic. Rationality isn't just the logical bit. Rationality is the outcome of selection from an adaptive repertoire that includes logic but is not limited to logic. Most emphatically, it is not logic alone.

Just how concerned should decision scientists be about the fact that human behavior often violates sensible logical axioms such as transitivity and independence? Here are more examples. Transitivity is simple: if A is preferred to B and B to C, then logic says that A must be preferred to C. But not infrequently, actual choices disagree. From a psychological point of view, the reason is pretty obvious: asked to choose between A and B, she chooses A; asked to choose between B and C, she chooses C; but then asked to choose between A and C, she may choose C. Her history—her *state* in technical terms—is different when asked to make the last choice than when asked to make the first. In A versus B, she has history X, whatever happened before the question was asked. But when asked C versus A, she has a history of A versus B and B versus C: her state is now different, so may her choice be different. No puzzle, if you are trying to understand the process, rather than thinking of the human subject as a time-independent logic device.

Independence is a bit more complicated. Look at the following table [table], which shows two pairs of gambles, A versus B and C versus D.[35] Boldface shows which expected value is generally preferred (again, there are always individual differences): B over A, but C over D. There are two things to notice about this pattern: First, most people pick the (very slightly) lower expected value in the

	probability			*Expected*
				Value
Gamble	0.33	0.01	0.66	
A	2500	0	*2400*	2409
B	2400	2400	*2400*	**2400**
C	2500	0	0	**825**
D	2400	2400	0	816

first pair but the (very slightly) higher EV in the second. Second, the only difference between these two pairs of gambles is that the payoff $0.66*2400$ (*italics*) has been added to both choices in A and B. Otherwise choice between A and B is the same as between C and D. Thus, most people's choices violate the independence assumption. The PT explanation is the primacy of the certain outcome (Gamble B), which describes the result but does not explain it. A causal account points to the repertoire of heuristics people have available to solve problems like this. A number of such heuristics have been proposed,[36] but expected utility remains part of most theories. The dominant approach is therefore an uncomfortable and probably unworkable compound of teleology and causality. It seems to me more likely that the explanation for people's modest irrationality is the complexity (which reflects an *internal constraint*) of the added probabilities which favors choice of the simpler option.

Violations of logical axioms like this should not consume so much mental energy, especially when the effects are typically small, as in this example. Behavior is not physics or predicate calculus. Human beings rarely maximize explicitly, following all the rules of optimality theory—especially when they must respond quickly to questions that may seem to allow more than one interpretation. They do not consider all the options in an unbiased, neutral, and history-independent way, then compute the costs and benefits and choose accordingly. They may misinterpret questions and use shortcuts. When their prior expectations are uncontrolled, they are rarely unanimous when responding to tricky choices. There is *no reason whatever* that human (and animal) choice should fit any set of simple logical axioms. The question is not 'are humans rational?' The question is 'how do people make decisions in a variety of contexts?" Utility-based economic theory is of very little use in answering the second question. As we confront what is increasingly perceived as a crisis in economics, looking for the underlying processes, the causes, of economic behavior will be of more use than continued preoccupation with utility maximization.

Notes

1 Smith, Smith, V. L. (1982) Microeconomic systems as an experimental science. *The American Economic Review*, 72(5), pp. 923–955. Vernon Smith shared the Economics

Nobel with Daniel Kahneman in 2002. Amos Tversky would have undoubtedly shared it also, but for his untimely death in 1996.

2 For an early collection of papers in this tradition, see Staddon, J. E. R. (Ed.) (1980) *Limits to action: The allocation of individual behavior*. New York: Academic Press.

3 Lewis, M. (2016) *The undoing project: A friendship that changed our minds*. New York: W. W. Norton.

4 Kahneman, D., and Tversky, A. (1979) Prospect theory: An analysis of decision under risk. *Econometrica*, 47(2), pp. 263–291.

5 See Chapter 3 for the many problems with the concept of statistical significance and the NHST method in general.

6 So-called *Giffen goods* are an exception: https://en.wikipedia.org/wiki/Giffen_good

7 Again, there are individual differences. For a miser, every dollar is worth the same, no matter how rich he becomes. The extravagant pay-inflation among top CEOs in recent years suggests that they also are exceptions to Bernoulli's principle—perhaps because their real reward is not the money itself, but their position relative to other big earners.

8 Expected value in this sense is just a mathematical quantity, probability times amount, that gives average outcome over many tries of a gamble. It has nothing to do with a person's *expectation*, in the psychological sense.

9 Stewart, J. B. (1991) *Den of thieves: The untold story of the men who plundered Wall Street and the chase that brought them down*. On June 5, 1986, Dennis Levine pleaded guilty to tax evasion, securities fraud, and perjury and became a government witness—but that is another kind of psychology!

10 Kahneman, D., and Tversky, A. (1979) Prospect theory: An analysis of decision under risk. *Econometrica*, 47(2), 263–291, p. 274.

11 Barberis, N. C. (2013) Thirty Years of prospect theory in economics: a review and assessment. *Journal of Economic Perspectives*, 27(1), 173–196.

12 https://en.wikipedia.org/wiki/Harry_Helson

13 Kahneman, D. (2011) *Thinking fast and slow*. New York: Farrar, Strauss, and Giroux, p. 344.

14 Both these examples involve two successive decisions and are not, therefore, directly comparable with one-step decisions.

15 See Edwards, W. (1954) The theory of decision making. *Psychological Bulletin*, 51(4), pp. 380-417.

16 This stricture seems to be ignored by many macroeconomic modelers. See Romer, P. (2016), above.

17 See, for example, Wilkinson, N., and Klaes, M. (2012) *An introduction to behavioural economics*, 2nd edition. Basingstoke, UK: Palgrave Macmillan, p. 163.

18 Kahneman, D. (2011), op. cit. p. 4.

19 Kahneman, D. (2011), op. cit. p. 55.

20 See also extensive discussion of consciousness in Staddon, J. (2014) *The new behaviorism*, 2nd edition. New York: Psychology Press.

21 Baddeley, A. (2007) *Working memory, thought, and action* (Vol. 45). OUP. Available at: www.oxfordscholarship.com/view/10.1093/acprof:oso/9780198528012.001.0001/acprof-9780198528012-chapter-1

22 Kahneman, D. (2011), op. cit. p. 88.

23 Observations like this led in 1908 to the Yerkes-Dodson law, which describes and up-and-down relation between motivation level and learning performance. Increased motivation is good, up to a point, after which performance declines. https://en.wikipedia.org/wiki/Yerkes%E2%80%93Dodson_law

24 Priming effects are weak. Priming experiments often fail to replicate. But there is little doubt that they do, sometimes, occur: www.ncbi.nlm.nih.gov/pubmed/27690509

25 Since the 'house' always wins, the very existence of Las Vegas is living testimony to the 'irrationality,' in rational-choice economic terms, of a substantial slice of humanity.

26 Minsky, M. (1986) *Society of Mind*. Available at: www.acad.bg/ebook/ml/Society%20 of%20Mind.pdf

27 www.britannica.com/topic/optimization

28 This idea could be tested experimentally by, for example marking the first peck with a change in some external stimulus—a key-light change, for example. If the change lasted until the next rewarded, pigeons would undoubtedly learn to peck at once after reward. But even if it was brief, they would probably wait less than without the stimulus change.

29 *Evolutionary fitness*—reproductive success, etc.—not gym-fitness.

30 Staddon, J. (2016) Theoretical behaviorism, economic theory, and choice. In *Economizing mind, 1870–2016: When economics and psychology met … or didn't*. Marina Bianchi and Neil De Marchi (2016) (Eds.) Durham, NC: Duke University Press, pp. 316–331.

31 In the published paper. Davis, D. G. S., and Staddon, J. E. R. (1990) Memory for reward in probabilistic choice: Markovian and non-Markovian properties. *Behaviour*, 114, pp. 37–64. The increases and decreases are computed by a Bayesian formula called the cumulative effects model, but many other law-of-effect rules have similar effects. See also: Staddon, J. (2017) Simply too many notes … A comment. *The Behavior Analyst*, https://doi.org/10.1007/s40614-017-0086-9

32 Horner, J. M., and Staddon, J. E. R. (1987) Probabilistic choice: A simple invariance. *Behavioural Processes*, 15, pp. 59–92. http://dukespace.lib.duke.edu/dspace/handle/10161/3231; Staddon, J. E. R., and Horner, J. M. (1989) Stochastic choice models: A comparison between Bush-Mosteller and a source-independent reward-following model. *Journal of the Experimental Analysis of Behavior*, 52, pp. 57–64.

33 See, for example, Elizabeth Kolbert's book review: That's what you think: Why reason and evidence won't change our minds. *New Yorker*, Feb. 27, 2017.

34 See, for example, Gigerenzer, G. (1991) How to make cognitive illusions disappear: beyond 'heuristics and biases.' *European Review of Social Psychology*, 2, pp. 83–115.

35 Kahneman and Tversky, Problem 1. This is an example of the Allais paradox: https:// en.wikipedia.org/wiki/Allais_paradox

36 Brandstätter, E., Gigerenzer, G., and Hertwig, R. (2006) The priority heuristic: Making choices without trade-offs. *Psychological Review*, 113(2), pp. 409–432.

7
'EFFICIENT' MARKETS

I take the market-efficiency hypothesis to be the simple statement that security prices fully reflect all available information.

Eugene Fama

Das ist nicht nur nicht richtig; es ist nicht einmal falsch!

Wolfgang Pauli

Economists are fond of terms borrowed from the physical sciences: friction, elasticity, force—and velocity, as in the *velocity of money*, for example. In physical science, these terms have appropriate *dimensions*, such as mass, length, and time. The dimensions of spatial velocity are distance over (divided by) time: velocity is a ratio. Money velocity does not, however, refer to the physical peregrinations of pounds or pesos. But it *is* a ratio: between a quantity of money and the number of times it has been exchanged per unit of time. For example, a $20 note paid to A on day one and spent on a meal in restaurant B, which then uses it to buy vegetables from supplier C the next day, will have a 'velocity' of 3 x 20/2 = $30 per day. For the economy as a whole, "[Velocity] can be thought of as the rate of turnover of the money supply—that is, the number of times one dollar is used to purchase final goods and services in GDP."[1] The general point is that money has to be spent to exert an economic effect, and therefore the rate at which it is spent may be critical. And the term does have the right dimensions for a velocity—amount/time.

Economics as Metaphor

Market efficiency is an old term for a market where things can be bought and sold quickly at a reasonable price. But this informal notion gave rise in the 1960s to a very formal theory. This chapter is about the puzzling rise, and surprisingly slow fall, of the *efficient market hypothesis* (EMH). The EMH is worth study from a methodological point of view because its conceptual basis, what it actually means, and how it may actually be tested, is so elusive. If we can understand the EMH, perhaps we can also learn something about the strengths and weaknesses of economic theory in general. Some of the discussion is unavoidably technical. I will try to make all as clear as possible.

Efficiency is another term from physical science. Like velocity, it is also a ratio: the efficiency of an engine is the ratio between physical work out divided by chemical or electrical energy in: miles per gallon or kilowatt hour. As we will see, *market efficiency* is much less easily defined. The seminal paper is: *Efficient capital markets: A review of theory and empirical work* (1970),[2] by Chicago economist Eugene Fama. Scholarly impact and a Nobel Prize in 2013 garnered 18,000 citations in 2017: not quite as popular as *prospect theory*, but a hit nevertheless. In what follows, I will argue that the notion of efficient market is conceptually flawed.[3] Any scientific model must be testable. But market efficiency is defined in a way that makes it impossible to measure and almost impossible to test. Efforts to do so have led to what look uncomfortably like wild-goose-chase attempts to capture an ever-mutating market process whose causal laws are not, and perhaps never will be, fully understood. To anticipate my conclusion: the problem with the EMH is that it is presented as a model but it is really an unfalsifiable *claim*, that the market cannot be beaten. Above-average profits cannot be made on the basis of publicly available information. But let's take the model idea seriously and see where it leads us.

Just how *is* market efficiency defined? Fama writes: "A market in which prices always 'fully reflect' available information is called 'efficient.'" This is the efficient market hypothesis. Fama calls it a "simple statement," but that is the one thing it is not. The definition differs from the way *efficiency* is used in physical science in several ways. First, it is not a ratio; second, it depends on something called *information* which can itself be defined in several ways; and finally, it doesn't yield a number. The terms 'available' and 'fully reflect' are also far from obvious.

Fama is aware that his 'simple' definition is in fact quite opaque:

> The definitional statement that in an efficient market prices 'fully reflect' available information is so general that it has no empirically testable implications. To make the model testable, the process of price formation must be specified in more detail. In essence we must define somewhat more exactly what is meant by the term 'fully reflect.'[4]

Well said. So, efficiency, which seems like a sensible and certainly a desirable property, depends on our understanding of exactly how prices are determined. In Fama's own words: "We can't test whether the market does what it is supposed to do unless we specify what it is supposed to do. In other words, we need an asset pricing model … that specifies the characteristics of rational expected asset returns in a market equilibrium…" I will return in a moment to the phrase "supposed to do" (according to whom? By what criteria? Etc.)

The term *efficiency* was in wide use before Fama's influential theory.[5] But efficiency as defined by the EMH is tricky, since it can only be evaluated by comparing actual market behavior with a model for how markets *should* behave. Let us accept for the moment that there are market equilibria, although I argued in Chapter 5 that they cannot be taken for granted. The price of something depends on how a thing is *valued*. Fama is apparently willing to ignore Jacob Viner's stricture against exploring the 'origin of value.' Fama is also breaking the prejudice against *forecasting* that I mentioned in the last chapter. These two defections are related, as we'll see in a moment.

The primary issue is: what determines price?[6] Can we predict (forecast) the price of a capital asset (or a portfolio of assets)—a bond, a stock, a stock option? How good is our prediction of price tomorrow, in a week, a month, or a year ahead? If we have a successful model, EMH fans must then ask: is it 'rational'? Does it 'fully reflect' available information about future earnings, growth, changes in technological, legal, business practice, etc? In other words, the essence of the EMH is a comparison between what the price *is* and what it *should be*. 'Should be' includes a rational estimate of *future* events, such as earnings and stock splits. In other words, it involves prediction, and not just prediction but 'rational' prediction.[7] (*Rational market hypothesis*[8] would be a more accurate label than efficient market hypothesis.) This looks like an almost impossible task. It assumes we know what the relevant information is and also exactly how price should react to it. It assumes that we know how the future *should* depend on the past, a very odd claim indeed.

Note that if, as the EMH requires, we know how to predict price, either actual price or what price should be, nothing is added by the term 'efficient,' other than a comforting label with historical precedent. And of course, if we do indeed know what price should be, and it isn't, we can make money by buying or selling, thus (possibly, but see below) restoring price to the proper level. Voilà, the efficient market rules!

Intrinsic Value

A simpler take on the elusive EMH, one that does not invoke 'information,' 'fully reflect' or rationality, seems to be Fama's early comment that: "[I]n an efficient market at any point in time the actual price of a security will be a good estimate of its *intrinsic value* [my emphasis]."[9] If we can discover the intrinsic value of an asset, we can compare it to its actual value. If the two are sufficiently similar, the EMH is supported.

Intrinsic-value EMH implies that there must be a negative feedback process, with the difference between actual price and intrinsic value constituting a sort of stabilizing error signal. Another way to put it, is that intrinsic value is what systems theory people would call an *attractor,* a stable equilibrium for every stock to which the market will return after any perturbation. In Fama's words: "In an efficient market ... the actions of the many competing participants should cause the actual price of a security to wander randomly about its intrinsic value."[10]

It seems to have escaped general notice that this claim is in direct contradiction to *random walk,* a popular model of an efficient market. A random walk has *no* self-regulatory properties. *Forced displacement,* an economic shock, just leads to further random deviations from the new baseline. There is no tendency to return to the original baseline. From this point of view random walk, far from supporting EMH, contradicts it. More on random walk in a moment.

Before going on to discuss technical details, it is worth saying that when an efficient market is described as 'rational use of relevant information' by market participants, and accepting the usual meaning of the term 'rational,' the EMH is pretty obvious nonsense. People buy stocks because they have always liked the company, or because they have used its products, or because of an effective advertising campaign. In other words, some people buy capital assets for reasons other than their financial prospects, and this fact has surely been known for a very long time: "Most stock market investors do not pay much attention to fundamental indicators of value" as Robert Shiller says.[11] In April 2017, for example, the market capitalization (total stock value) of Elon Musk's charismatic Tesla company (US market share for cars 0.2%) was about the same as the value of General Motors (market share 17.2%). Tesla seems to be the stock-market equivalent of torn jeans or mechanical wristwatches. Yet financial prospects are the only independent (predictive) variables in EMH models.

"Not a problem" say EMH defenders. There are enough 'rational' investors to keep prices within bounds and maintain market efficiency. The 'irrationals' are just random noise around the 'rational' mean.

Accordingly, Fama defends the EMH by noting that market deviations from its predictions tend to be random, equally often too large, and too small:

> First, an efficient market generates categories of events that individually suggest that prices over-react to information. But in an efficient market, apparent underreaction will be about as frequent as overreaction. If anomalies split randomly between underreaction and overreaction, they are consistent with market efficiency. We shall see that a roughly even split between apparent overreaction and underreaction is a good description of the menu of existing anomalies.[12]

This is a curious defense. After all, a running average of daily prices would also show daily deviations equally often above and below the average. But it would hardly constitute an adequate model of the pricing process.

Even more curious is the following: "Following the standard scientific rule, however, market efficiency can only be replaced by a better specific model of price formation, itself potentially rejectable by empirical tests." While strictly true—a model should only be replaced by a better one—Fama's claim is irrelevant to the main scientific issue, which is *disproof*. It is not true in science that "it takes a theory to beat a theory," as more than one economist has claimed. A theory may be discarded simply because it is wrong. The phlogiston theory, which predicted that a substance loses weight after being burned, was disproved by the finding that burning (oxidization) often makes things heavier. It was not necessary to discover oxygen or invent the atomic theory to show that the phlogiston theory is false. The existence of the aether was disproved by the famous Michelson–Morley experiment, which found that the speed of light is the same in all directions, despite the movement of the earth. Relativity theory, which explained their result, came much later. A model that purports to predict prices can be disproved when it fails to do so. An immediate replacement is not required.

One more comment on market efficiency before I get to an actual model. If P is price and V is 'intrinsic value,' presumably efficiency would amount to a measure of the difference between price and value. The smaller the difference, the higher the efficiency. Would positive deviations, $P > V$, count the same as negative ones: $P < V$? If so, a measure of market efficiency could be $E = 1/|P-V|$ (where $||$ denotes absolute value) or even $E = 1/(P-V)^2$ (either form[13] guarantees that our efficiency measure $E \geq 0$).

Most efficiency measures are limited to $0 < E < 1$: efficiency is output/input and output cannot exceed input. (*Miles-per-gallon* looks like an exception, but not if both input and output are expressed in energy units.) It is not possible to limit in this way a measure of market efficiency measured in terms of price and intrinsic value, because there are no limits to P or V. Price may exceed value and vice versa, so P/V need not be less than one.

This is all academic, of course. No quantitative measure of efficiency exists. Lord Kelvin's dictum applies: even fans of EMH agree that it cannot provide us with any quantitative measure of efficiency. And if it did, the important achievement would be estimating V, the 'intrinsic value' of an asset; 'efficiency' would just be a ratio in which V is involved. In fact, there is no conceivable way of estimating V with confidence. Indeed, there may be no such thing as intrinsic value. Sometimes V is related to measurable things like (projected) future earnings or the growth potential of a startup. There have been a few semi-successful attempts to find a formula that works, as we will see. I discuss one possibility, the Capital Asset Pricing Model (CAPM) below. But sometimes it is just Tesla and torn jeans. In practice, V is simply what we are willing to pay; it is the same as price.

This uncertainty has not prevented at least one eminent economist from claiming to know intrinsic value, at least approximately: "Fischer Black, in his 1984 presidential address before the American Finance Association, offered a new definition of market efficiency: He redefined an 'efficient market' as 'one in which price is within a factor of 2 of value, i.e., the price is more than half of value and less than twice value ... By this definition, I think almost all markets are efficient almost all of the time."[14] Nice to have his opinion, but how could he possibly know? Black's statement is about historical volatility not value.

Models

In the absence of inside information, changes, especially short-term changes, in the price of a capital asset are usually unpredictable. If they were predictable—if there was *regression to the mean*,[15] for example—money could be made algorithmically by buying undervalued assets. Traders would take advantage. Given the elementary realities of supply and demand, predictability will *arbitraged away* as the jargon goes. Then, price changes will become unpredictable again. Any successful price-predicting algorithm, once it is widely used, is *self-canceling*. It may work at first, but as soon as traders catch on, a once-effective formula will lose its power. Market processes by themselves will ensure that asset prices are unpredictable.

Random Walk

This apparently commonsense argument is only partly true. But it led many years ago[16] to the *random walk* model of stock-price movement. There are several versions of random walk (also, rather confusingly, called *martingales* and *fair games*). One of the simplest is that price changes in each discrete time step by a small random amount, *p*, a random number between −1 and +1, so that increases and decreases are equally likely. Current price is independent of previous prices; a random walk has no memory. RWM means that each price change is independent of preceding ones, just as each successive coin toss is independent of the previous series of heads and tails. RWM-driven prices are therefore completely unpredictable.

Real markets show a few deviations from this pattern. Prices occasionally show 'momentum' where rises tend to follow rises and falls to follow falls. The time period over which this trend will continue is unpredictable, so that the 'random' nature of price variation is preserved. There seems to be a slow upward trend to stock prices.

Asset prices also show a feature that makes them quite different from, say, coin tosses, accident statistics or even the weather:

> [T]here is *better forecastability ... of speculative asset returns for longer time horizons*. This accords with longstanding advice that investors should be patient, that they cannot expect to see solid returns over short time intervals. But this is just the opposite of what one would expect in weather forecasting where experts can forecast tomorrow's temperature fairly well but can hardly forecast a year from tomorrow's temperature[17]

> *[my emphasis]*

These properties, not to mention common-sense understanding of how business enterprises work, suggest that slow, long-term processes are involved in the success of an enterprise and the price of its stock. Hence, short-term, 'local' predictive models are unlikely to work well. Time is important: what matters in a financial market is surely the time rate of return. Predictability in the short run would allow for a high rate of return to trading; long term predictability, not so much. Hence short-term predictability will be the first to be eliminated by market forces.

Reinforcement learning (see Chapter 5: operant conditioning) presents an analogous picture. In operant conditioning, as in the market, the aim is to maximize return: solve the problem as fast as possible and get the reward. But RL organisms rarely follow a marginalist maximizing strategy. They learn via variation and selection. As we saw in Chapter 6, successful models for operant choice are not local, but take remote history into account. The similarity to market behavior is just this: in both cases, behavior is much more variable, hence much harder to predict, earlier, before much selection has occurred, than it is later when fully adapted to the situation. Perhaps this inversion—low predictability after a short time, more predictability if you wait—is characteristic of all outcome-driven processes, such as operant learning, running a business, or investing in one.

The RWM can be tested in several ways. A hypothetical record can be generated simply by daily coin flips. Can a professional stock 'charter' detect a trend in this 'stock'? About as well as he can in a real stock, apparently. People tend to see patterns even in randomly generated data.[18] One can look at real stock prices to see if there are any sequential dependencies: is a rise more likely after a previous rise? And similarly for a fall? In other words, is each up or down shift independent of preceding ones?

On all these tests, real stock-price fluctuations usually, if not invariably, look pretty random. There are a few exceptions. A real random walk deviates from its starting point at a rate proportional to the square root of time. Often, but not always, true for stock prices; sometimes the variability of stock prices increases more over time than RW predicts.[19] A large sample of stock prices can show a slow drift, usually up, but sometimes down. A set of truly random walks would not show that. Random walk may go below zero; real stocks cannot do that. Most importantly, real stocks occasionally show very large changes. When you hear talk of an 'astonishing 6-sigma [standard deviation] price movement,' something that, apparently, should not occur even over geological periods of time, the reference point is a standard random walk. If prices move according to the RWM, such large shifts should essentially never occur. What is wrong of course, is the RW assumption. Despite these relatively infrequent exceptions, real asset-price variations, especially short-term variations, are often close enough to a random walk to keep the theory alive.

Does that mean that stock prices are *really* random? No, for three reasons. First, information other than that represented by price and profit history is important.

When oil giant BP suffered the Deepwater Horizon oil spill in the Gulf of Mexico, its stock dropped precipitately. So-called event studies, which I discuss in a moment, are another example of nonrandom behavior of asset prices. External events can have a big effect. The EMH accepts effects of public information so events like Deepwater Horizon do not invalidate it. But inside information, not represented in its current price or publicly available recent history, can also affect the future price of a security in a nonrandom way.

Second, *random* is a statement of ignorance. Proving that something is random is in effect proving a negative, which is impossible. To say that something is random is to say only that you don't know what the rule is. You don't know what formula or algorithm determines the series of numbers you're looking at, or even whether such an algorithm exists. For proof, just look at the (pseudo)random-number generator on your computer or the digits of the transcendental number π. In both cases, every digit is completely determined, so the series is not random at all. Yet by all the standard tests these series do not deviate much from randomness.[20] When you hear the word random, think *inexplicable.*

Third, as I mentioned earlier, random walk is not self-correcting.[21] If a price is displaced—by an erroneous company report, for example—RW would simply continue varying from the new baseline. RW-determined price, unlike real prices, would show no tendency to return to its previous level. On the other hand, the EMH allows price to be affected by information. In the example, the initial displacement can be traced to a rumor. If price returns to its former level, perhaps the recovery can be traced to diffusion of the fact that the rumor is false? This is the chameleonic nature of the EMH: If prices vary randomly, EMH is confirmed. If they do not, it must be the effect of new information.

Although some influential writers have equated RWM and EMH,[22] the two have some incompatible features. RW implies complete unpredictability, whereas the EMH—at least in the form advocated by Fama—is tied to the idea of 'rational prediction.' Nevertheless, RW behavior is offered by some as proof that the market is efficient, even though random behavior, by definition, follows no identifiable rule. But first a comment on another conundrum.

The Joint Hypothesis Problem

When I first encountered Eugene Fama's forceful and persuasive accounts of the EMH, I was puzzled by his frequent references to what he calls the *joint hypothesis problem.* It took me a while to see that it was in fact a nonproblem, just a by-product of the epistemological flaws of the EMH itself. Let me go through the argument.

In his 2013 Nobel essay, Fama writes:

> Tests of efficiency basically test whether the properties of expected returns implied by the assumed model of market equilibrium are observed in

actual returns. If the tests reject, we don't know whether the problem is an inefficient market or a bad model of market equilibrium. This is the *joint hypothesis problem* emphasized in Fama (1970).[23]

[my italics]

And later:

> If the test fails, we don't know whether the problem is a bad model of market equilibrium … or an inefficient market that overlooks information in setting prices … This is the joint hypothesis problem.

The joint-hypothesis problem is a false opposition. The scientific issue is not whether the market is efficient or not, but whether your model works or not. You test a market model you think represents 'efficiency.' It fails: bad model. That's all. Was your model 'efficient'—who knows? Who (actually) cares, since the model is wrong. Repeat with a new model: This time the model works, it predicts future price. Is the model 'rational'? Depends on what you mean by 'rational,' a question more for philosophers than scientists. Likewise, for 'efficient.' Why should a scientist care, since the model works (at least for a while!). For science, all that matters is whether a model survives test. Whether a successful model is 'efficient' or not is a matter of personal taste, since the term has no testable consequences.

Stability and the EMH

A final comment on stability: "Any successful trading algorithm, once it is widely used, is *self-canceling*," i.e., provides *negative* (stabilizing) *feedback*. What is wrong with this as a justification for EMH? The answer is "not *any* but *some*." Not all trading algorithms are self-canceling. Here's one that is: an inverse stop-loss-type rule. 'Buy, if price is below X; sell if it is above Y (Y >> X).' If most traders follow this rule, price will presumably stabilize between X and Y and the rule will no longer work. The rule is stabilizing *negative feedback*. On the other hand, consider this momentum-type rule: 'Buy if price today is greater than price yesterday; sell if the reverse.' Pretty obviously, this rule is highly destabilizing *positive feedback*. Given enough compliant traders and a little initial variability, price will either tumble or escalate into a bubble. A momentum rule is destabilizing.

Two conclusions follow from this dichotomy. First, to the extent that traders use self-canceling algorithms, stock prices will be unpredictable moment-to-moment, and the market will be stable and appear to be 'efficient.' Second, market stability depends on the laws that govern buying and selling by individual traders. These laws are not known and may never be known. Hence market stability, equilibrium, cannot ever be taken for granted.

Data

Asset-price models are evaluated by using either historical (longitudinal) price data on individual assets (stocks, bonds, etc.) or by comparing data from many assets across a brief interval of time (cross-sectional data). The method is *inductive*: past market history (measured in various ways) is used as a guide to future market behavior. All the swans I know are white, *ergo* the next swan I see will be white. The usual assumption is that the process underlying market models remains essentially the same across whatever period is under study. A model that works for the first half of the period will also work for the second half. This assumption is usually false.

In an examination of the effect of dividends on stock prices, Robert Shiller asks: "Should we take the latest ten years real dividend growth as a guide to the future, rather than the last 30 years or some other interval?" Applying the same dividend-discount model to both periods, and seeking to predict the same subsequent period, Shiller finds very different results.[24] So, either the predictive model he uses is wrong, or the process being modeled in the last ten years is different from the process in the preceding twenty.

The fact that measures based on stock prices will vary depending upon the time period chosen is now well accepted: "Beta calculated with ten years of data is different from beta calculated with ten months of data. Neither is right or wrong—it depends totally on the rationale of the analyst" opines an online financial dictionary.[25] (More on *beta* in a moment.)

Eugene Fama in a 2007 interview[26] tells a story that makes the same point nicely:

> When I was at Tufts, I was working for a professor who had a stock market forecasting service. My job was to devise rules for predicting the market, and I was very good at it. But he was a pretty good statistician. He always told me to set some data aside so I could test [the rules] out of sample. *And they never worked out of sample.*
>
> *[my emphasis]*

Is there a defined time period within which we can be sure that the process we are attempting to model is stable? Perhaps. Can we identify it? Usually no. The reason is that all these models are statistical, not causal. They use past correlations, sometimes quite complex correlations, to predict future prices. They do not, because they cannot, incorporate the actual process by which prices are determined.

The actual process of course involves thousands of individual market participants making up their own minds about what to buy and sell and subject from decade to decade to the whims of fashion, 'animal spirits' and the advances of financial science. Modeling this multitude in detail is obviously impossible. Instead, the tacit assumption is that something like the gas laws operate. Millions of individual gas molecules, each individually capricious and unpredictable, yet behave in aggregate in a lawful way. Is there something like a Boyle's Law of markets? Are all these market participants, individually unpredictable, nevertheless collectively lawful in way that can be captured by mathematical models?[27]

Perhaps, but there is a critical difference between people and gas molecules. The molecules move independently; their paths are not correlated. But people talk to one another, they read about trends, they are exposed to advertising, propaganda, and all manner of rumors. Individual buy and sell decisions are not independent of one another. The behavior of market actors is correlated and the amount of correlation changes over time, high during 'panics' and 'bubbles,' lower at other times. Traders are probably *never* entirely independent of one another. It is quite possible to imagine a model that takes account of feedbacks among agents, but as far as I'm aware no one has yet proposed one.

Event Studies

The EMH, with its emphasis on information, has led to some interesting, quasi-experimental tests—a rarity in economics where the scope for real experiment is limited. I discuss the rather technical example of so-called *event studies,* not because the results are particularly surprising, but as an illustration of a quasi-experimental method used by Fama and his collaborators to arrive at a conclusion they believe is consistent with EMH.

Event studies are the closest that economics can come to a simple AB experiment. The logic is straightforward: compare the performance of a security before and after an event such as an earnings report, a stock split, or some macroeconomic event such as a currency shift or a new regulation. Ideally, the timing of the event should be completely independent of the variable being measured. Relevant questions are: (1) What is the stimulus, the event? (2) How is stock value measured? (3) What is effect of the event? (4) How was the stimulus controlled: what determined its time of occurrence? Much thought has been given to (1), (2) and (3), and much less to (4). I discuss these four factors separately in a moment.

In many cases, the results of these 'field experiments' are straightforward. For example, the picture shows the effect of a quarterly earnings announcement (stimulus event) on the difference between expected return on the investment and the actual return (CAR: cumulative abnormal return, also cumulative average residual, explained more fully in a moment), before and after the announcement.[28] The CAR is cumulative, so a flat line at any level indicates 'no change' from time period to time period. The pre- and post- period was 20 days and the stocks were 30 companies in the Dow-Jones index and the sample period was January 1989–December 1993. Each announcement was put in to one of three categories: good, bad, and neutral, and the average CAR for each group plotted separately.

As you can see, the results are pretty much as one might expect: The good-news group rises slowly before the announcement, which is followed by a sharp rise which is sustained. The bad-news group shows almost the opposite: a slow

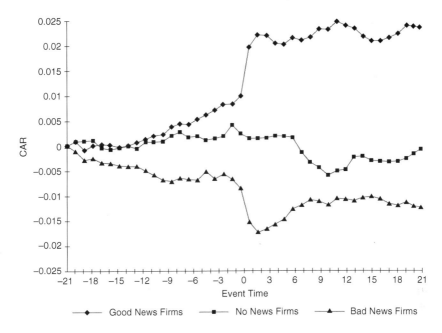

Source: MacKinlay, A. C. (1997) Event studies in economics and finance. *Journal of Economic Literature*, 35, pp. 13–39, figure 2a.

decline pre-announcement, followed by a sharp decline and then some recovery. The no-news group, as expected, shows essentially no change before and after the announcement.

The CAR is partly under the control of the company, which therefore has some ability to anticipate the known release date of each quarterly report. It may decide to push earnings into one quarter rather than another, for example. In other words, the event is not a totally *independent* variable, as it should be for the study to be a real experiment. Nevertheless, the results suggest that this minor confound had negligible effects.

This independence is not so true for the event study that has been used to support the EMH. I look next at the famous Fama, Fisher, Jensen, and Roll (FFJR, 1969) study.[29] Eugene Fama places much weight on this analysis of a slightly more complex company event than an earnings report—a stock split.

1 Nature of the stimulus

A stock split involves no change in a company's assets or profitability; it adds no new information. But the timing of a split is under the company's control and does reflect factors related to stock price. Companies split their stock—two-for-one, for example—when they are doing well and their individual stock has risen well above average for their industry and possibly to a level that deters small investors. Deciding exactly *when* to split a stock is not simple: current price, the price of competing stocks, and the company's present and predicted fortune, are all involved in the decision. Again, a stock split involves no change in a company's assets or profitability; it adds no new information.

2   How to assess stock value

Fama *et al.*'s analysis begins with a survey of month-by-month data on returns of 692 stocks (a much larger sample than the MacKinlay study) over the years 1926–1960. Fama *et al.* fitted a linear regression[30] to all these returns:

$$R_{it} = a_i + b_i R_{Mt} + e_{it},$$

estimating a and b for each stock, where R_{it} is the return on stock i for month t, R_{Mt} is the market return, plus a 'noise' term, expected to have zero mean, termed the *residual*, e_{it}, which is the difference between individual data and the sample mean.[31] This regression represents a hypothesis: that the month-by-month return on a stock can be predicted by a linear sum of two things: the return of the market as a whole, R_{Mt} (weighted by constant b_j), and a constant a_j; e_{it}, the residual, is what is left over, the unpredicted variation. If the stock just varies in line with the market, the mean cumulative average residual (CAR) will be 'noise,' with a mean of zero. A consistently nonzero residual indicates an effect of the event being studied, a stock value significantly above or below that predicted by previous market data.

Since FFJR were going to use the regression results to predict the effect of events (in this case, stock splits), they wanted to exclude from the original regression those regions of data where changes might be expected. First, they identified some 940 instances of a stock splitting in their 33-year sample of 692 stocks. They then excluded, for the moment, a time period around the time of each split, using a criterion based on a regression result from the entire dataset: "This criterion caused exclusion of fifteen months before the split for all securities and fifteen months after the splits followed by dividend decreases." The authors add in a footnote "Admittedly the exclusion criterion is arbitrary." But they repeated the analysis without exclusion and got much the same results. To assess the effect of a stock split on returns they plotted residuals, i.e., the discrepancy between the average residuals in the neighborhood of a stock split and the average from the data as a whole.

3   What is the effect of a stock split?

In effect, data—residuals—from all the stock splits were superimposed, lined up by the time of the split, and then averaged. The result is shown in the picture: The deviation from expected value of the residuals slowly increases in advance of the

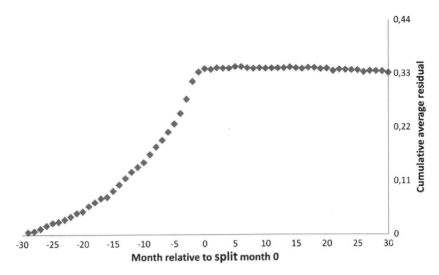

Cumulative average residuals in the months surrounding a split.

Source: Fama, Fisher, Jensen, and Roll (1969).

split but does not change afterwards. The graph shows no after-effect of a stock split. The curve is flat showing that the residuals after the 940 splits average out to zero (CAR stays constant) after a split. Which suggests, says Fama, that the split by itself adds no new information to the pre-split increase in stock price. This proves (Fama concludes) that all the information about company performance was factored in to the stock price before the split. The change in advance of the split reflects (FFJR claim) anticipation of future divided increases; and indeed, when they carried out a similar analysis for dividend increases and decreases they found a similar result: residuals increase in advance of both an increase and a decrease, remained steady after an increase, but decreased slightly following a dividend decrease.

But a naïve experimenter (me, for example) looking at these data might well conclude precisely the opposite. The rise in CAR before the split ceases immediately afterwards. *That* is the effect of the split.

4   What controls the stimulus?

The timing of a quarterly earnings report, the stimulus event in the previous analysis, is fixed and not under the control of the company issuing the report. Not so for a stock split, which is totally under the control of the company and can occur at any time. The simplest hypothesis to explain the results in the picture, therefore, is that companies split their stock when they judge that it has hit a plateau. The fact that the CAR does not change after a split implies that, on average, they are correct.

A puzzle for the EMH is that no further increments in CAR occur after a split despite the fact that "75% of the companies that split their stocks continue to experience good times (witnessed by subsequent dividend growth rates larger than those of the market as a whole)." Fama takes the lack of change in CAR after a split as support for the EMH: "[O]n average, all the implications of a split for the future performance of a company are incorporated in stock prices in the months leading up to the split, with no further reaction thereafter—exactly the prediction of market efficiency." But if a stock split adds negligible new information, as most observers seem to agree, why did not CAR continue to rise for those companies after the split? Perhaps '75% of companies' is not sufficient to affect the average? We don't know.

Dividends are intimately related to the 'real' value of a stock so form a big part of any 'rational' predictive equation. How do stock prices based on expected future dividends compare with what might be expected from EMH? Not too well it turns out. The problem is that predicted aggregate price follows a relatively smooth upward curve (e.g., from 1870 to 1980), but actual aggregate price does not: "it is consistent with the [EMH] … that there can be sudden shifts in price when there is important new information about subtle changes in trend. But it would seem that important new information should be something that occurs only rarely, given the smooth nature of dividends." says EMH critic Robert Shiller.[32] And later: "One very basic thing that is learned from [these data] is that the model that people essentially know the future, a model that is often suggested as an approximation, is wildly wrong in all periods."[33] If price is seriously dependent on dividends, then since dividends change only gradually, there is just too much variation in price. As we saw earlier, even a completely deterministic process may show sudden changes in the absence of any external input. But this conclusion only emerges if we pay attention not just to equilibria but also to 'laws of change.'

Perhaps stock price depends on more than dividends, so the divergence in amount of variability between dividends and stock price is large not because EMH is wrong, but because dividends are not the only thing that determine stock price? What else might be involved? I look next at models that try to link market measures with changes in stock price.

CAPM

In finance, unlike the prospect-theory experiments, money equals utility; Bernoulli does not apply. Marginal utility does not diminish with amount. A fundamental equation of expected-utility theory is that net value = $U.p$. That is, the actual value of a choice is equal to the utility of the outcome times its probability of occurrence. The Chapter 6 discussion of prospect theory provided many examples. Probability, p, is termed *risk*. In an asset-market context, p would be the chance that a given security would drop to zero value—if the parent company goes broke, for example. The *Capital Asset Pricing Model*, which, like the marginal-value theorem I discussed in Chapter 5, was proposed independently by several theorists,[34] is perhaps the simplest way to apply this very basic idea to financial assets. It has since become a sort of handy Mr. QuikPrice for securities analysts. CAPM was the first choice as an answer to Fama's quest for: "an asset pricing model … that specifies the characteristics of rational expected asset returns in a market equilibrium. …" Despite its ubiquity, the CAPM is flawed in many ways, as we will see.

As I mentioned earlier, many years ago Chicago economist Frank Knight distinguished between what he called *risk* and *uncertainty*. The distinction is real but is often ignored. *Risk* is quantifiable. In the terms I used earlier, in Chapter 2, risk is the well-defined number you can assign to an outcome when you know in full the sample space. With two dice the sample space contains 36 possibilities, all equally likely: 11, 12, 13, …, 66. If you bet on two sixes, you know *exactly* your chance of winning (1/36) and your chance of losing (35/36); similarly for roulette or Bingo. All the prospect theory examples I discussed earlier involved risk in this sense. All the possibilities and their probabilities are known.

Uncertainty is much trickier. There are obvious examples: no one could have given the odds, or perhaps even imagined the possibility,[35] of the smartphone the day after Alexander Graham Bell's announcement. No one could have given odds on the horrific train of events following the assassination of an obscure Archduke in Sarajevo in 1914. 'Being at the right (or wrong) place at the right time' refers to events that are uncertain, have no quantifiable odds. The chance of winning a military engagement involves uncertainty rather than quantifiable risk. 'Market risk,' the chance that the market will behave in the future precisely as it has done in the past, is really market uncertainty.

There are two unknowns in the expected-utility equation $V = U.p$: utility and risk. There is actually uncertainty in both terms: the firm may go broke (that would be V and U both zero) and future returns may not be the same as returns in the past; they might better or worse (an error in U). But, so says CAPM, let us just take past returns, r_A, as a measure of present utility. What about risk? Unfortunately, stock risk is Knightian and not quantifiable. We never know the sample space, the set of possible future values for a stock and their probabilities. So, whenever you see the word 'risk' in discussions of financial matters, read *uncertainty* (and be afraid, be very afraid!).

Nevertheless, investors, and their advisers, need some more or less objective way to value a security. So, the CAPM inventors morphed the uncertainties of real stock risk into the familiar form suggested by expected utility. The sample space for a stock value is unknown, but its *volatility* (price variability trade-to-trade or day-to-day—the time period is arbitrary) is easily measured. Why not take volatility as a *proxy*: the greater the volatility, the greater the risk? The CAPM, and many other market risk models, use volatility as a way to measure risk. But the risk of asset A must be judged in relation to the risk associated with the market as a whole. In other words, the investor is interested in A's volatility in relation to the overall volatility of the market.

β

Which brings us to something called *beta* (β). β is a number greater than zero that represents the volatility-based *systematic risk*, that is, the risk associated with a given asset, independently of general market risk. The dependent variable, the thing investors usually want to maximize, is *return*,[36] suitably adjusted for risk. Two volatilities can help estimate risk: volatility of return on the asset itself and average volatility for the market as a whole. Obviously, we would like risk-adjusted return for the asset under consideration to be greater than the risk-adjusted return for the market as a whole—at the very least.

To derive β, just consider the two relevant factors: market return volatility, σ_M^2, and the covariance[37] between the asset return, r_A, and the market return r_M: $\text{cov}(r_A, r_M)$. We want β to be zero when the market and the asset are uncorrelated and 1 when they are perfectly correlated, which can be accomplished by setting:

$$\beta - \text{cov}\left(r_A, r_M\right)/\sigma_M^2.$$

It is easy to see that β has the desired properties. If the asset return is totally uncorrelated with the market return, $\text{cov}(r_A, r_M) = 0$, there is no systematic risk and $\beta = 0$. Conversely, if asset return is perfectly correlated with market return, then $\text{cov}(r_A, r_M) = \sigma_M^2$ and $\beta > 1$, large systematic risk.

So (the argument goes) now we have a measure of systematic risk we can just plug it into a commonsense comparison with the risk-free rate (a Treasury bond, say), r_F, giving the CAPM formula:

$$E_A = r_F + \beta_A\left(r_A - r_M\right),$$

where E_A is a measure of the value of the asset. Only if $E_A > r_F$ is the asset worth purchasing. The asset with the highest E_A is preferable to all others. The term in parentheses is called the *risk premium*, since it is the amount by which the asset return must exceed the risk-free return to balance β.

This little equation looks as if it provides an estimate of actual asset value, but of course it cannot, because the risk numbers are uncertainties, not actual probabilities. What it does do is allow stock analysts to rank order securities in terms of their likely desirability as investments.

Like many predictive devices applied to asset prices, the CAPM is self-canceling. "The golden age of the model is … brief," comments Fama in his Nobel essay. The processes assumed by models like CAPM are approximations that hold for a while, but become increasingly poor guides the more they are exploited by traders.

Various improvements to CAPM were suggested. Perhaps the most influential was the Fama-French[38] '3-factor model.' Using cross-sectional data, the model derives the return on a stock or a portfolio as a linear function of two independent variables in addition to the risk-free return used by CAPM: size of security and book-to-market equity ratio. In 2015, the authors added two more factors. Others have suggested nonlinear models incorporating variables like trading volume.[39]

The way that financial modeling proceeds can be summarized roughly as follows:

1 Assume that return, y, on a given stock portfolio can be predicted by a set of 'publicly available' variables obtained from historical data: $x_i...x_n$:

$$y = F(x_i...x_n),$$

2 For example, suppose F is linear, as in most existing models, we can write (1) as

$$Y = a_1 x_1 + a_2 x_2 + ... + a_n x_n.$$

CAPM used basically two variables (x values, $n = 2$), Fama–French three and later five.

3 Parameters $a_1 - a_n$ are estimated from cross-sectional data and the resulting equation is used to predict returns over some future time period. But ...

4 Each version has its day, then after while ceases to work well. What is wrong?

What is wrong is what is 'at risk' in any inductive process: that all swans may *not* be white. A predictive procedure that works now may cease to work in the future. Or, more technically, even if a linear model is appropriate, there is absolutely no guarantee that all those parameters, $a_1...a_n$ will remain constant indefinitely. Indeed, as I pointed out earlier (Stability and the EMH) given the complex feedbacks inherent in markets, there is every likelihood that parameters will change unpredictably from epoch to epoch.

Market models are *statistical fits* rather than hypotheses about actual causes. But unless a model really captures the causal process that underlies movement in asset prices, any inductive approximation, any curve-fitting, is unlikely to remain true indefinitely. Weather forecasting, with which market forecasting is often compared, doesn't suffer from this problem, for two reasons. First, its equations are (simplified) versions of atmospheric physics. They are approximations to real causation. Second, the weather, unlike a financial market, is unaffected by a forecast. Weather forecasts have no effect on the weather, but financial forecasts affect the behavior of market participants, hence affect the market.

A purely inductive, statistical strategy may be the best that can be done in finance. It is useful as a sort of quantitative market research. But it is not adequate as science: *C'est magnifique, mais ce n'est pas le guerre!*

Whither the Efficient Market?

What then shall be the fate of the EMH? It has failed multiple tests. Yet even people who have convincingly shown it to be wrong still seem to treat it as some kind of reference point. Many eminent economists are reluctant to abandon the efficient market hypothesis. Unpredictable 'nonequilibrium' behavior is treated as a sort of puzzling discrepancy, a temporary deviation from the market essence, which is efficiency.[40] One reason the EMH is so resilient is that it is so hard to define precisely. It reminds me of Winston Churchill's comment about the old Soviet Union: "a riddle wrapped in a mystery inside an enigma." It's hard to hit a shape-shifting target. Another reason it persists is that calling a market 'efficient' is great PR: who could be against efficiency?

Nevertheless, the fact is that there is no such thing as an efficient market. The idea is conceptually and empirically flawed. To the extent that clear predictions can be derived from it, if efficiency is interpreted as random walk, for example, the data don't fit. In other forms, the EMH is simply incoherent.

But I believe that allowing the EMH to survive even on life support is a mistake. As Wolfgang Pauli famously said about another theory: "It is not even false!" The EMH is not true; it is not false; it is not defined; it is not definable. It amounts to the claim that you cannot—cannot!—beat the market with public knowledge, which is an unfalsifiable claim.

Bad science makes for bad policy. The end-of-the-rainbow search for market efficiency has two ill effects. First, it provides a convenient rationale for every kind of opaque 'market-making.' Highly leveraged options and derivatives "may look like nonsense to you, John Doe, but these things increase market efficiency!" Well maybe, since we don't really know what market efficiency is. On the other hand, these complex instruments very often contribute to market *instability*, something which is both coherent and measurable. The shadowy presence of an unmeasurable market efficiency distracts attention from market stability, which is both measurable and demonstrably important to the economy as a whole. Stability, well-defined and attainable, is surely a better guide for regulators than efficiency, which is neither.

Notes

1 The definition according to the Federal Reserve Bank.
2 Fama, E. (1970) Efficient capital markets: A review of theory and empirical work. *Journal of Finance,* 25(2), pp. 383–417. Also: Two Pillars of Asset Pricing, Nobel Prize Lecture, December 8, 2013. There is another, and much better defined, kind of market efficiency called *Pareto efficiency* (after Vilfredo Pareto (1848–1923), Italian engineer and economist). A market is Pareto-efficient (*optimal* would be better term) if it allocates resources in such way that any re-allocation will make someone worse off. https://en.wikipedia.org/wiki/Pareto_efficiency
3 See also *The Malign Hand*, Chapter 10 for a brief critique of efficient market theory.
4 Fama, E. (1970) p. 383.

5 See, for example, Sewell, M. (2011) History of the efficient market hypothesis. *UCL Research Note* RN/11/04. www.e-m-h.org/

6 A popularizer of the efficient-market idea proposes a version apparently simpler than Fama's: "a definition of efficient financial markets [is] that they do not allow investors to earn above-average returns without accepting above-average risks." But Burton Malkiel still requires some model for *risk*, the most difficult thing to assess in pricing an asset. Malkiel. B.: The efficient market hypothesis and its critics. www.princeton.edu/ceps/workingpapers/91malkiel.pdf

7 There seems to be a little confusion here. Fama's view of efficient markets seems to require rationality, but Malkiel claims "Markets can be efficient in [the sense that investors cannot make above-average, risk-free returns] even if they sometimes make errors in valuation … Markets can be efficient even if many market participants are quite irrational. Markets can be efficient even if stock prices exhibit greater volatility than can apparently be explained by fundamentals such as earnings and dividends." So, for Malkiel, the market does not have to be rational to be efficient. What, then, would make a market *inefficient* for Malkiel?

8 See Justin Fox *The myth of the rational market* (2009) Harper-Collins, e-books, for an interesting discussion of the EMH and its history.

9 Fama, E. (1965) *Random walks in stock-market prices. Selected Papers, no. 16.* Chicago, IL: University of Chicago Press. *Intrinsic value* seems to be the same thing as *utility* (see Chapter 6). One of the problems of social science is that the number of technical terms much exceeds the number of concepts to which they are supposed to refer.

10 Fama, E. (1965) *Beware "affirming the consequent": EMH implies random-walk-like variation in stock price. But RW stock variation does not imply EMH.*

11 Shiller, Nobel address.

12 Fama, E. (1998) Market efficiency, long-term returns, and behavioral finance. *Journal of Financial Economics,* 49, pp. 283–306.

13 A small constant would need to be added to each denominator to limit E to a finite value when $P = V$.

14 Robert Shiller, Nobel address.

15 Regression to the mean, first discovered by Francis Galton, is simply the fact that a sample from a normally distributed process that happens to be at one end or the other of the bell curve (Chapter 3), is likely to be followed by a larger or smaller sample, closer to the average.

16 See Holt, J. (2013) A Random Walk with Louis Bachelier. *New York Review of Books,* 60(16), pp. 63–64. Available at: www.nybooks.com/articles/2013/10/24/random-walk-louis-bachelier/, for a quick review. Random walk is also sometime called *Brownian motion* after Scottish botanist Robert Brown who, in the 19th century, looking at tiny pollen grains suspended in water, noticed that they seemed to constantly jiggle about. Albert Einstein in 1905 showed that the movement was due to random buffeting by the thermal motion of water molecules. For videos see https://en.wikipedia.org/wiki/Brownian_motion and www.youtube.com/watch?v=jLQ66ytMa9I

17 Robert Shiller, Nobel address. Of course, climate-change advocates claim to be able to forecast trends acting over decades or centuries better than conventional weather forecasters can predict over a month or so, which is another exception to the short-term-predictions-are-better rule. See Curry, J. A., and Webster, P. J. (2011) Climate science and the uncertainty monster. *American Meteorological Society,* December, pp. 1667–1682 for an excellent critical review of this controversial issue. Physicist Robert Brown has an eloquent blog on the theoretical issues: https://wattsupwiththat.com/2014/10/06/real-science-debates-are-not-rare/

18 www.sciencedirect.com/science/article/pii/019688589190029I

19 Lo, A. (1999) *A non-random walk down Wall Street.* Princeton, NJ: Princeton University Press.

20 www.huffingtonpost.com/david-h-bailey/are-the-digits-of-pi-random_b_3085725. html

21 https://en.wikipedia.org/wiki/Ornstein%E2%80%93Uhlenbeck_process

22 Malkiel, B. G. (2003) The efficient market hypothesis and its critics. *The Journal of Economic Perspectives,* 17(1), pp. 59–82. Available at: www.princeton.edu/ceps/ workingpapers/91malkiel.pdf

23 This last sentence is puzzling because I could find no reference to 'joint hypothesis' in the 1970 paper.

24 Shiller, Nobel address, figure 1.

25 www.investinganswers.com/financial-dictionary/stock-valuation/capital-asset-pricing-model-capm-1125

26 www.minneapolisfed.org/publications/the-region/interview-with-eugene-fama

27 Agent-based market models are reviewed in http://pmc.polytechnique.fr/pagesperso/ dg/offer/market2.pdf and www.researchgate.net/publication/259384100_Agent-Based_Macroeconomic_Modeling_and_Policy_Analysis_The_EuraceUnibi_Model

28 MacKinlay, A. C. (1997) Event studies in economics and finance. *Journal of Economic Literature.* 35(1), pp. 13–39. Figure 2a.

29 Fama, E. F., Fisher, L., Jensen, M. C., and Roll, R. (1969) The adjustment of stock prices to new information. *International Economic Review*, 10, pp. 1–21.

30 Fama, E. F. Nobel address, Eq. 5.

31 In the original 1969 paper, these returns are scaled logarithmically. There is no mention of scale in Fama's Nobel lecture.

32 Shiller, Nobel address, p. 469.

33 Robert Shiller, Nobel address, pp. 469–470, figure 1.

34 https://en.wikipedia.org/wiki/Capital_asset_pricing_model

35 C.S. Lewis actually came close in his science-fiction story *The Machine Stops,* (1909).

36 There are many ways to measure return on an asset. See, for example: https://en. wikipedia.org/wiki/Return_on_assets

37 Not quite the same as correlation, see https://en.wikipedia.org/wiki/Covariance_ and_correlation

38 Fama, E. F., and French, K. R. (1992) The cross-section of expected stock returns. *The Journal of Finance*, 47(2), pp. 427–465.

39 E.g., Shi, L., Podobnik, B., and Fenu, A. (2017) Coherent preferences and price reference point jumps in stock market. *3rd International Workshop on "Financial Markets and Nonlinear Dynamics* (FMND), 2017. www.fmnd.fr/

40 See, for example, Lo, A. W. (2017) *Adaptive markets: Financial evolution at the speed of thought.* Princeton University Press. Kindle Edition.

8
SUMMING UP

To kill an error is as good a service as, and sometimes even better than, the establishing of a new truth or fact.

Charles Darwin[1]

[W]e live in an age [when] there is a considerable amount of intellectual tyranny in the name of science.

Richard Feynman[2]

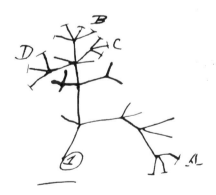

Science is like evolution in at least one respect: it has developed as a sort of branching process. Social science, especially, has become ever more specialized. Part of this specialization is the natural result of growth. But a larger part is probably due to professionalization, the fact that science has gone from being a vocation for a privileged minority to a career for a large number, many of whom must earn their wages by demonstrating progress. The result is less and less interaction across areas and the development in each subfield of conventions about appropriate data, type of theory, valid methods of test, and legitimate problems of study, that serve both scientific and sociological ends. They are little subject to scrutiny from outside. They are tools to achieve truth, yes, but only in ways that are not incompatible with career advancement. Consequently, some areas of science have drifted off into irrelevance, embracing untestable or incomprehensible theories and, if they do experiments at all, doing them in a fragmentary way that distracts rather than explains.

In every case, the visible apparatus of science—controlled experiment, mathematically rigorous theory—is present. Yet, even in areas where is experiment is possible, like social and personality psychology, we almost never see the kind of time-consuming trial-and-error sequence of experiments that leads to some general principle or important phenomenon—like the discovery of oxygen, Newton's *Opticks*, or Piaget's 'stages.' In social science as a whole, all the appearances are there. All that is lacking is substance.

In social and personality psychology, for example, the publication norm seems to be one or a few experiments of three kinds: clever demonstrations of common sense, equally clever demonstrations congenial to academic fashion, or cute demonstrations showing something strikingly counterintuitive. The highly successful fraudulent studies by the Dutch researcher Diederik Stapel I mentioned in Chapter 4 show examples of all three types. But he, like the field, tended to favor issues that play to fashion.

Stereotype Threat

One more example: another play to fashion that made a big hit is something called 'stereotype threat.' The seminal paper[3] garnered more than 6,000 citations by 2016. The study seemed to show that the apparent inferiority of African-American to white students at solving an intellectual task (questions from a Graduate Record Examination study guide) was due almost entirely to self-perception. Because black students are stereotyped as poor test-takers, they will do worse if they think they are taking a test rather than simply solving a problem. When the GRE task was presented in a neutral way, as problem-solving rather than an IQ-type test, the black-white difference apparently vanished. The blacks were as good as the whites. The media take-away was that stereotype threat alone is responsible for the substantial (one standard deviation) difference between white and black SAT scores.

But Steele and Aronson compared not raw scores, but scores 'corrected' for the participants' SAT scores. They did find a small effect of the way the task was presented, but in the wrong direction. Stereotype threat made black students perform worse; absence of threat did not make them perform equal to whites. ST increased the difference between black and white students; absence of threat left the difference unaltered:

> [R]ather than showing that eliminating threat eliminates the large score gap on standardized tests, the research actually shows … [that] absent stereotype threat, the African American–White difference is *just what one would expect* based on the African American–White difference in SAT scores, whereas in the presence of stereotype threat, the difference is *larger* than would be expected based on the difference in SAT scores[4]
>
> *[my italics]*

Nevertheless, such was the world's relief at finding that the black-white intellectual disparity could apparently be traced to prejudice, rather than to biology or culture, the ST paper was misinterpreted for many years.

Economic Models

Denied easy experiment, economists resort to models. Models seem to play a different role in economics than theory in biology or the physical sciences. Newton's equation, that the force attracting two bodies to one another is proportional to the product of their masses divided by the square of the distance between them— $F = Gm_1m_2/r^2$—is trivial mathematics. But each term is measurable and well-defined. Relativity apart, there is no other gravity model. Most economic models are functional rather than causal and not tied to any specific real-world situation: 'Utility is maximized under the following conditions…' If a situation fits the conditions, then the model applies. If its assumptions hold, each model is true as a tautology; the models are in effect theorems.

Economics, boasts many such models. Some economists regard this diversity as a strength: "The diversity of models in economics is the necessary counterpart to the flexibility of the social world. Different social settings require different models."[5] The skill of the theoretical economist then is not so much in the model as in its application. The models themselves belong not to science but to applied mathematics. What is left of economics becomes not science but art. The economist's role, as economist not mathematician, becomes his ability to discern a match, not perfect perhaps, but adequate, between a real-world situation and an appropriate model. The 'science' of economics apparently resides not in verifiable knowledge but in the intuition of economists. All else is math.

There are three ways an economic model can fail. Only two of them are much discussed by economists:

1   The theorist may simply misperceive the situation. A familiar example is the Israeli day-care experiment in which parents were fined for picking up their

kids late. Supply-demand theory predicts that late pick-ups, previously cost-free, should diminish once they incur a cost. Instead, they increased, because late parents were willing to pay a charge, but less willing to fail an obligation.

2 Critical assumptions may not be true. For example, many situational constraints must hold for the Black-Scholes model for pricing options to work: no possibility of a riskless profit, no transaction costs, prices vary according to a random walk with drift, etc. In reality, these limitations may not hold or may be impossible to verify: Was there perfect competition? Do prices follow a random walk, really? Economists are well aware of this limitation on models, but also frequently ignore it.

3 And finally, a problem that is rarely discussed: conflation of causal and functional (teleological) models. Most economic models are functional not causal. They find an optimal solution to a maximization problem. They are outcome-based. Such a model may fail to match reality *even if all its assumptions are fulfilled*. The examples I gave in Chapters 5 and 6 show the reason: an organism may behave optimally, may satisfy an optimality model, without *optimizing*. Rational behavior can happen without rational thought, just as a dog can catch a Frisbee without knowing the laws of physics.[6] Failure to distinguish between explicit optimizing—comparing marginal utilities, for example—and achieving an optimum in some other, and more fallible, way, is an error committed by almost all economic models. If causes, like mass and distance, have been accurately identified, a causal model will always work. But a functional model, no matter how appropriate to the situation, will only work if the underlying causal processes are adequate. Sometimes, like the matching-law example in Chapter 5, the simple reward-following process that subjects seem to use is adequate to yield optimal choice. But sometimes, when the reinforcement schedule is more complex, it is not and behavior is suboptimal. In the real human economy, choices are made by countless individuals according to rules that are not only different from one person to another, but are still only dimly understood. The causes of market choices can never be totally mapped. So functional models may fail even if all the necessary conditions are satisfied.

Yet optimality models remain attractive, for several reasons. First, they often yield fixed, equilibrium solutions, which have for years been the starting point for economic theory. Paul Samuelson's 1941 paper, for example, looks at supply and demand curves, whose intersection gives the market equilibrium. All that remains is to measure—'identify'—the two curves to get a complete theory:

> Thus, in the simplest case of a partial-equilibrium market for a single commodity, the two independent relations of supply and demand, each drawn up with other prices and institutional data being taken as given, determine by their intersection the equilibrium quantities of the unknown price and

quantity sold. If no more than this could be said, the economist would be truly vulnerable to the jibe that he is only a parrot taught to say 'supply and demand'... In order for the analysis to be useful it must provide information concerning the way in which our equilibrium quantities will change as a result of changes in the parameters taken as independent data.

So, 'laws of change' seem to be required. Nevertheless, Samuelson's analysis takes supply and demand curves, which are functional not causal, as his starting point. He treats them as causes, which they are not. He never comes to grips with the real problem, the underlying real-time dynamics. The efforts since Samuelson's paper to identify actual supply and demand curves, and their lack of success, seems to indicate that this is not the right question.

Tinbergen's Four Questions

It is both a strength and a weakness of maximization theory that it works so powerfully as a unifier. Having a single answer is especially attractive to economists, who must often make policy recommendations: "After all, while there is only one way to be perfectly rational, there are an infinite number of ways to be irrational." in the words of economist and *New York Times* columnist, Paul Krugman.[7] I first came to appreciate the unifying power of optimality theory many years ago when behavioral ecologists were beginning to point out, often very elegantly in quantitative models like the marginal-value theorem discussed in Chapter 5, how various aspects of animal behavior were optimal, how they maximized Darwinian fitness.

Unlike economists, ethologists (students of animal behavior) and behavioral ecologists can do realistic experiments. They are not limited to functional models; indeed, they can often test them experimentally. Instead, the field was structured around four broad questions (of which function was only one) proposed by the great ethologist Niko Tinbergen.[8]

The four questions are first, *function*: how does a characteristic contribute to evolutionary fitness? Using proxies for fitness such as feeding rate or number of offspring, economic optimality models can be usefully applied to biological problems. The second question is *causation*, which is the prime concern in the physical sciences, can be answered experimentally by biologists, but is rather neglected by economists. Tinbergen's last two questions concern *development* and *evolutionary history*. Both are relevant to economics, since the history and evolution of a practice has a bearing on its function and effectiveness. But these factors are rarely discussed in economics textbooks.

Behavioral ecologists were able to bring together many disparate phenomena under the umbrella of Darwinian fitness optimization. I was personally delighted to discover that behavior on many schedules of reinforcement could be unified by a theory of reward-rate maximization. For example, empirical functions relating molar (average-rate) measures of responding (low-cost) and reward (high-value) have distinctive shapes, depending on the reward schedule. On random-interval schedules, for example, the functions are mostly positive and negatively accelerated: response rate increases with reward rate. But on random-ratio schedules they are mostly declining: response rate decreases as reward rate increases. These differences are predicted by models that balance the cost of responding against the benefits of reward.[9]

But it soon became apparent to me that optimization has scientific limitations. For one thing, it is too flexible. As a prominent biologist pointed out, you can always find a way to define any behavior as optimal by changing the constraints or the measure of utility. In fact, optimization is most informative when it fails, because then the investigator must return to the causal level. To understand why the organism is not behaving optimally—'rationally'—you must understand what it is doing, the actual process involved.

The causal analysis of choice, unlike the rational/optimal choice theory, is not a unifier. Unfortunately, choice and decision making depend on what I have been calling *variation*, as I described in connection with prospect theory in Chapter 6. Choice behavior involves learning, which is a process of behavioral variation and selection. The processes, strategies etc., offered up by variation are the material on which selection must operate. Behavioral variation comprises many 'silent processes' and the outcome of their competition is often uncertain. Biological evolution itself is intrinsically unpredictable. If human learning and decision-making works along similar lines, we are too often left with 'don't know' and not rarely 'can't know.' Conclusions like these are not helpful if part of an economist's *raison d'être* is to advise policy makers.

Optimization, on the other hand, offers certainty. It is attractive because it gives relatively simple answers in the form of equilibrium solutions. In practical matters, 'know' always beats 'don't know.' In addition, economists are interested (as Jacob Viner might have said) in the economy, in what is happening 'out there' not in the buyer or seller's head. That is *psychology*, after all. If psychology can be ignored, optimization rules.[10]

There is a certain elegance in the mathematical proofs of optimization theory. But the field has, I fear, been rather carried away. Simple ideas, clearly expressed, are not enough. Critics like Paul Romer argue that infatuation with this kind of elegance has got out of hand, rendering even simple statements almost incomprehensible. It seems that mathematics must intrude, even if it obscures.

Here is an example from Eugene Fama's 'efficient market' Nobel address: The verbal statement to be explicated is "Tests of efficiency basically test whether the properties of expected returns implied by the assumed model of market equilibrium are observed in actual returns." A paragraph of formalism follows:

A bit of notation makes the point precise. Suppose time is discreet [sic], and P_{t+1} is the vector of payoffs at time $t + 1$ (prices plus dividends and interest payments) on the assets available at t. Suppose $f(P_{t+1} | \Theta_{tm})$ is the joint distribution of asset payoffs at $t + 1$ implied by the time t information set Θ_{tm} used in the market to set P_t, the vector of equilibrium prices for assets at time t. Finally, suppose $f(P_{t+1}| \Theta_t)$ is the distribution of payoffs implied by all information available at t, Θt; or more pertinently, $f(P_{t+1}| \Theta_t)$ is the distribution from which prices at $t + 1$ will be drawn. The market efficiency hypothesis that prices at t reflect all available information is

$$f\left(P_{t+1}\mid \Theta_{tm}\right)=f\left(P_{t+1}\mid \Theta_t\right). \tag{1}$$

Does labeling 'information' as "Θ" really 'make the point precise?' Does anyone think that the formalism and the equation adds anything whatever to the last sentence of the quote? Are we any nearer to understanding what 'reflect information' actually means? Are we nearer—and this is promised by so much math—to a quantitative measure of efficiency?

What about nonsocial science—biomedicine, epidemiology? There are two problems. First, as I discussed, experiment is often impossible, for ethical reasons, or impractical, because of long time-delays between cause and effect. The result is that statements such as "2004 study showed that power plant impacts exceeded 24,000 deaths a year..." abound in the press. Where is the proof that these deaths (out of an annual total 2.4 million) were uniquely caused by (not 'correlated with' or 'linked to') not just pollution but pollution from power plants? A slightly more specific report, from MIT, is this: "Study: Air pollution causes 200,000 early deaths each year in the United States New MIT study finds vehicle emissions are the biggest contributor to these premature deaths."[11] I say, 'slightly more specific' because the headline at least draws attention to the fact that it is *years lost* that is critical, since life is fatal for us all. But the unavoidable disease of epidemiology remains: these data are all correlations, not the results either of experiment (ethically impossible) or basic science that has uncovered the precise physiological mechanisms by which pollution acts.

Epidemiology is a source of hypotheses, not causes. Yet the media, and many scientists themselves, have largely erased the difference between the two. Little has changed since E. B. Wilson pointed out the problem more than 70 years ago. These correlational revelations almost qualify as 'fake news'; but the topics attract attention even if the claims are dodgy. So, into the headlines they go.

The philosopher Bertrand Russell many years ago wrote a wonderful book called *Human knowledge: its scope and limits.* My aim in this book has been to make a contribution to Russell's great effort through the medium of examples—of phenomena explained by science or scientific papers that made an impact or make a point. If this discussion can help readers fairly to assess the blizzard of scientific and pseudo-scientific information that rains on the public every day, it will have been of use.

Notes

1 Letter to Stephen Wilson, 1879.
2 True in 1966, even more so now. http://profizgl.lu.lv/pluginfile.php/32795/mod_resource/content/0/WHAT_IS_SCIENCE_by_R.Feynman_1966.pdf
3 Steele, C., and Aronson, J. (1995) Stereotype threat and the intellectual test performance of African-Americans. *Journal of Personality and Social Psychology*, 69(5), pp. 797–811.
4 Sackett P. R., Hardison C. M., and Cullen M. J. (2004) On interpreting stereotype threat as accounting for African American-White differences on cognitive tests. *American Psychologist.*, 59(1), pp. 7–13. The authors discuss at length the misrepresentation of the ST paper by some textbooks, as well as PBS's *Frontline* and other media stories.
5 Rodrik, D. (2016) *Economics rules: The rights and wrongs of the dismal science.* New York: W. W. Norton & Company. Kindle Edition. p. 5.
6 See Haldane, A. (2012) 'The dog and the Frisbee,' for a discussion of this issue by an economist: www.bis.org/review/r120905a.pdf
7 As paraphrased in a biographical article by Larissa MacFarqhar, *New Yorker*, March 1, 2010.
8 NT's older brother was Jan Tinbergen, Nobel-winning economist.
9 Staddon, J. E. R. (1979) Operant behavior as adaptation to constraint. *Journal of Experimental Psychology: General*, 108, pp. 48–67.
10 Well, some economists are interested: Bianchi, M., and De Marchi, N. (Eds.) (2016) *Economizing mind, 1870–2016: When economics and psychology met … or didn't.* Durham, NC: Duke University Press.
11 Jennifer Chu, MIT News Office, August 29, 2013.

INDEX

Note: Et seq. indicates a topic is discussed in pages following the given page.